STAIRWAY TO A DREAM

A journey from blunders to wisdom

Alexander Markitanov

Translated by
Fedor Markvardt

ABOUT THE AUTHOR

ALEXANDER MARKITANOV is a psychologist, life coach, and a writer. He discovered a unique method of how effectively avoid different problems we encounter in life. The main areas of his work are psychology of health and psychology of relationships.

At the age of 30, Alexander Markitanov got hurt in a motorcycle accident. Doctors sentenced him to a life in a wheelchair, but he overturned this verdict. He started to look for a way out of his trauma by seeking ancient knowledge of self-healing methods from different sources. He worked hard and eventually cured himself. It was the beginning of his study of Eastern medicine and self-healing methods. He went to India, Tibet, and China where he studied healing systems based on Ayuverdic practices, yoga, and Tibetan teachings.

When Alexander Markitanov achieved a level of mastery, he became a teacher of the Institute of Human Self-Healing. He left his job and found his calling in spreading the knowledge he acquired. Alexander founded the School of Expansion of Consciousness and Vesta School of Women's Wisdom to promote healthy lifestyle and happiness. These days, Alexander Markitanov conducts various trainings in Russia, Ukraine, and Canada.

Please visit website for more information: http://markitanov.com

FriesenPress

Suite 300 - 990 Fort St
Victoria, BC, V8V 3K2
Canada

www.friesenpress.com

First Edition — 2016

Translated by Fedor Markvardt.

ISBN
978-1-4602-8452-0 (Hardcover)
978-1-4602-8453-7 (Paperback)
978-1-4602-8454-4 (eBook)

1. SELF-HELP, PERSONAL GROWTH, HAPPINESS

Distributed to the trade by The Ingram Book Company

LOVE is the formula of life.

CHAPTER 1

Not too much time had passed since I had met a man named Victor at the airport, on the way to Delhi. He was the man whom I considered to be my first mentor.

This most colourful of characters told me a downright shocking tale that was both compelling and invaluable. Through the tale I realized that I could control my life, and more importantly I figured out how to do it. The method was simple – get rid of all of my "mental trash". The more "mental trash" a person has, the more a person depends on circumstances. By the time of our meeting, I was more than ready to get rid of the nuisance of the unnecessary internal cargo that I had carried in my mind and heart throughout my life. I realized that it was time for the reconstruction of my consciousness. I saw clearly the direction I should go - towards development, turning inwards to a spiritual search. I audited my own life's values, and I rearranged their priorities which allowed me to gain a certain feeling of security from my imperfections. Only after I made these internal rearrangements, I did feel a sense of order in my mind and comfort in my soul. After I had a feeling that something important was going to happen and I felt both anxious as well as curious about what my future held. And indeed, what followed was a series of amazing and unforgettable events.

First, I made a rule for myself: *I had to see any problem or failure as an opportunity to practice helpful thoughts and get rid of the useless and non-constructive ones.* If I experienced failure, broken plans, or obstacles to my goals, I would not allow myself to worry. Instead, I tried quickly to understand why it had happened, and then I went about correcting the situation. Even if I did not understand or correct a situation right away, then *I would accept the situation and learn an appropriate lesson from it.* I knew for sure that my change in attitude would create a positive situation. I got rid of the fear and uncertainty. In space they used to occupy I put peace of mind, comfort, and enjoyment of life. I really enjoyed my new state! A few weeks later after that fateful meeting, I had already started to receive pleasant surprises from my simple internal changes.

I was happy and I felt that I had already hit my limit of change. I prayed to God to strengthen my new position and help me stay with it because it was good and could not be any better, or so I thought. In fact, that one fateful meeting was just the beginning of the intensive lessons, events, and meetings which opened new horizons for me that I had not even dreamed of.

In general, life began to gain momentum and at some point I started to meet "unfortunate" people in the same situation I had been in. I saw people who were in crisis or who were at a dead end. They sought for answers to the important questions of life. At that time, when I spoke with a person who had some problems, I was teaching and giving that person advice in order to help them, but I was still learning myself. I found myself thinking of how I was becoming like Victor (my mentor) to others and the people who I was helping - were like me but before I'd met Victor. At the airport I had seen Victor as spiritually strong and confident, so I had listened to him attentively.

I had thought, "He is amazing! It would be great if I could deal with my problems, make the right decisions, and control my life as he does. If I could learn how to apply Victor's strategy to my life, I could achieve all of my goals."

What was I doing with my life? My business required lots of time, so it used up all my energy. My combat classes I attended less frequently and I found that I needed another way to calm my nerves and lessen my stress. I found a simple and reliable way – drinking.

My friend Arthur would say, "It is necessary to periodically install the gasket between the hemispheres to prevent a short circuit". The "gasket" was, of course, the alcohol. Then I began to understand that alcohol couldn't really solve anything. I began to pay attention to older people who earned a lot of money and "relaxed" after work by drinking alcohol. Many of them had paid for their wealth with their physical and/or mental health. Their pursuit of wealth had been achieved, but their expected joy hadn't manifested. I concluded that people lose something on the way to their goals. Perhaps, with the growth of wealth, personal growth, and their upwards climb in social status there was also a need to grow something else. To find that "*something*" it is very important to build up the value of one's priorities. I had to figure out what came first in my life and what didn't. A proper structure of values in life allows you to abandon secondary goals in favour of the main ones. This strategy, over time, allows a person to gain wisdom.

I realized how valuable my time was. I wanted to use it for the development of my body, mind, soul, and spirit. I started to read a lot and took control of my thoughts, emotions, words, and actions. I started to take combat classes again, but instead of violently beating on the punching bag, I began to comprehend the philosophy of the classical masters of the martial arts. I realized that a deep insight into the philosophy helped my progress, not only in the fight competitions, but in life as well. As a result, I started to become more confident and my business improved.

People started seeking advice from me because they could see the changes in my life and I gave my advice freely because I could help them. I thought, "Perhaps there is a law in the universe - if I learn the truth I must share it with others. That is why the universe introduces the seekers to those who have found answers. Those who need support will find it. If so, then that is why I met Victor."

I read clever books, exercised, and trained my mind and spirit. Eventually, I began to think that I would reached Victor's level in spiritual development. I thought that since I had become a mentor myself, how great it would be to meet Victor again, so he could see the changes in me for which he had been a catalyst.

As you know, if a person desires something from a pure heart, then it will happen. While I was in Moscow, I happened across an announcement for a presentation of his new book. "That's interesting!" I thought. "He wrote a book, but he said that he didn't have any skill at writing. Good for him! He probably worked hard developing his abilities and discovered a hidden talent. I was no longer surprised after that realization."

I did not miss the opportunity to go to the presentation of his book. From the first few moments of Victor's interaction with the audience, I realized that my thoughts of reaching my mentor in spiritual development were no more than an illusion - a manifestation of my own pride. That is how it happens: I lost a bit of control over my thoughts and made a fool of myself. As the Eastern sages say, "He got trapped by his own ego." I forgot one of the laws of the universe - *never compare yourself to anybody and never compare anybody with yourself.*

I was ashamed of my thoughts, but I remembered Victor's words about the law of a self-support, so I quickly let go of the feeling of guilt. I was still learning and I wasn't perfect. *Even if you made a mistake - remember it, try not to repeat it, and move on supporting yourself.* This principle supports internal growth.

By listening to Victor's speech, it was clear to me that he had obtained more skills. He became even more likeable and it seemed that he emitted an inner light. Victor was a great example of how people can live and continue to transform over time.

I felt a strong desire to cultivate my own best qualities, inner strength, and virtue. I wanted to commit to actively improving myself and I understood that there was no limit to what I could accomplish if I did.

After the presentation we had a nice chat. I was curious to find out what had happened in his life and I wasn't disappointed as he spoke about the new places he had gone and the new people that he had met. It was extremely interesting.

"Victor, tell me how often you come across foolish people like me?" I asked, and I thought to myself, "Two hours ago, I saw myself as a wise man and now I see myself in the other extreme – the exact opposite." I noted that I could laugh at myself in this situation and I was happy that in my illusions of self-sufficiency I did not go too far - I had stopped in time. I pointed out that lack of experience in controlling my thoughts to myself- I had been captured by my ego. It was a good lesson for me and the nicest thing about it - it was painless. All these thoughts flashed in my mind seconds before Victor, with a wide smile, answered my question.

"Self-irony undoubtedly is a useful quality but do not be carried away by putting such a stigma on yourself. We meet many different people in our life time, but that conversation at the airport I remember well. Not everyone will listen to sermons from a fellow traveler. In my opinion, you have a wonderful quality – you are open to new things. When a person has an interest and a desire to learn it's the start to a fascinating conversation. Usually, I talk to people in my office. My door is always open for consultations for anyone who is willing to overcome their problems in life. A lot of, people come for help, but they are closed off, so often I meet strong opposition."

"But you said that you have a technical profession?"

"I finished my psychology degree, and I changed my occupation. I decided to create a spiritual-educational centre, so I started with the office."

"That's great! A lot of people probably come to you with some amusing cases. Can you tell me about some of them?"

"Yes, there are many interesting cases. One of the first meetings I remember very well."

* * *

Victor began his story and he had my full attention.

One day a young man named Edward came to see me. He entered the room, sat down, and stared silently at the floor. He didn't know where to start.

"So, what can I do for you my friend?" I asked him.

He was then looking into my eyes silently and we studied each other for a while. I saw that he did not trust me and did not dare to speak frankly with a stranger, but he had apparently gotten tired of his problems and had come for help despite that. He looked down again, sighed, and spoke...

"Since I was young, people who were cruel filled me with disgust and anger." Edward said, still looking at the floor. "I used to beat them mercilessly. Growing up I began to realize that people are capable of being mean; playing mind games and dirty tricks. That's because they aren't good people but it causes me to have negative emotions as well as aggression which isn't good. I don't want these emotions because I feel like they are destroying me. Tell me what to do when people do such ugly, dishonorable, and vile things? The actions of these dishonest people are a threat to my interests. Should I, as before, use force to chastise them, even though it keeps happening? Where do they come from? I have a feeling that I am doing something wrong but I don't know what exactly. It seems like I'm at a dead end."

"Let's see how we can deal with it together. When we come across bad behaviours in human nature, we might experience the feelings of condemnation, anger, resentment, and fear. The stronger a stimulus, the stronger emotional the reaction to it. You're right. Negative emotions and bad feelings are destructive. They destroy the flesh and clog up the inner world of the person. I think that, in order to break the cycle, you must fundamentally change your perceptions of the world. Try this idea: *Don't worry, don't be responsible for the actions of others, and don't judge anybody.* An adherence to this principle as a rule will allow you to avoid negative feelings and condemnation. Reactions to a stimulus are born of your mind and they are different depending how a mind adjusts to a stimulus. You should affirm in your mind the following: *Let each person deal with and be responsible for their actions.* Put your thoughts in the right direction. Right means useful, allowing you to continue on your own journey. So, set a task for yourself: *Change a state of condemnation, anger, and fear to a state where you can control your response to an incident. Try to find something positive in a situation.* Do you understand?"

"Find something positive in people being mean?"

"That is what you think. That is your perception of an event. After all, these guys are helping you to become a truly strong and joyful person."

"What? The bastards are helping me?! Bad things from bad people help me? I think you're pulling my leg, Mr. Psychologist! How is that possible?"

"We are talking about the new model. Unfortunately, in high school you and I did not learn the laws of the universe. But even a superficial knowledge of these laws can help to avoid a lot of mistakes."

"What laws? Are you going to reveal the secret of how spaceships travel across the universe?" The man muttered.

"Something like that. I suggest you take a look at things from a different angle - from the height of the stars." I calmly answered ignoring the frustration and the aggression of my client. "First of all, from the perspective of the Creator - there are no good or bad people and there are no good and bad circumstances. There are just people and circumstances. We judge other people and circumstances depending on our attitude. Judging is no more than an interpretation of a situation by a mind. A mind relies on the accumulated knowledge of a person which is usually full of delusions. One event and different people will see it in different way and, sometimes, in a completely opposite way. This means, judging can easily be subjective, and often that is the case."

"So, what do I do?"

"It's very simple - *avoid judging*. This is one of the laws of the universe."

"How do you know, and how can you prove to me that there is such a law?"

"Dear Edward, I really want to help you, but I have no desire to prove anything to you. I can only share with you my observations and my thoughts. I can tell you my opinion on your problem and if you want, you can accept it. That's how I can help you. So, do you agree to work with me?"

"Well, if you're an expert, you have to prove your point."

"You and I have different points of view on how to be a professional. Different views are normal. Do you support the idea that in a dispute the truth comes to light?"

"Yes, of course. How else would it?"

"I have a different opinion. If your position is to have proof, it means that you are focusing on opposing new information. You've united with your concerns about asking someone else to help solve your problems. This is a position of disadvantage because if you intend to find an effective way in solving your problems, you have to be open to it. Do you get it?"

"Yes, I think so. Not entirely, but I agree with your point. What do you suggest?"

"I suggest you actively participate in solving your problems. This means that you have to find the way by yourself. Just see how the information that I give you can be useful. My life and professional position is not to argue or impose. No one can be forced to become happy, wealthy, or healthy. What do you think?"

"Hmm ... well, reasonable. Perhaps I accept."

"Then let's go back to the law - do not judge anybody. Do you think God is judging and acting selectively? Does God divide his own children as good or bad? Does God describe the events that happen to us as good or bad?"

"I don't know. Possibly. What does it change?"

"Excellent question! That's the point. Nothing changes. If God judges, then give God this case, so it will allow you to stop judging. If God doesn't judge, you shouldn't think that you understand God's work better than God does. Therefore, I suggest you to study the laws of the universe. Search and find God in you. God does not judge. God loves all people equally. God created a mechanism according to which everything moves and takes place in the world and the Universe. If you comprehend this mechanism, you will find that if someone commits evil acts for their own personal desires, then that person will reap what he or she created."

"Perhaps you make it too complicated? Maybe everything happens spontaneously. Why nobody knows about these laws or mechanisms?"

"You can say that, by interviewing at least one thousand people that no one knows and it still won't be accurate. Have you done this research?"

"No I haven't."

"Then you can only speak for yourself. Expressing an opinion on behalf of everybody is an attempt to evade responsibility for what has been said by dividing the responsibility with everyone; but without even asking anyone. It's a trick of the mind and one of the most common and absurd habits. It's wiser to speak on your own behalf. To express your views on your own behalf is one of the indicators that a person is taking responsibility for their own life. A person who speaks on his/her own behalf is more responsible for their own statements. In this case, it is necessary for you to better understand faster your thoughts. How do you like this?"

"Well, OK."

"Even if someone doesn't know the laws of the universe, those laws won't change and they will continue to exist and work. As an example - radio waves exist whether we see them or not. One way or the other, if we want to learn something, we need to take some action. However, in the intricacies of life it is not so simple and clear. In this case you must be observant and study the issue. This is part of the action. So, you believe that everything happens by chance, right? Then I will try to answer your question, 'Who is right?'."

"Who?"

"The strongest is always right. It is an established fact, even if the strongest is supporting an incorrect idea."

"Where is the justice in that?"

"Right here. If the good wants to win, then it must be strong not allowing itself to whine or to be weak. And evil is a stimulus for the good to become strong. Good will lose if it's weak. If good wants to win, it must be strong to be right. Do you think you are strong?

"Absolutely!"

"Do you agree that the good is creative and the evil is destructive?"

"Everybody knows it."

"I ask you to speak only for yourself. Practice shows that "everybody" may not agree with you. Do you agree that love this is the formula of life?"

"Yes, I agree."

"What is the opposite of love?"

"Hatred."

"I share your point of view. I just want to add that evil begets evil and cruelty breeds cruelty."

"Sure."

"As you told me, you want to live without evil and a bad people who do bad things, right?

"Well, yes. Though, I wish there were less of those people."

"Now look what comes out. If you have hatred to people, you're begetting the evil! You assume that the world shouldn't be the way it is, but the way you want by creating illusions. By heaving this idea, you feed your illusions. Illusions, in a collision with a reality generate internal conflict and consume your energy. The fact that you consider yourself a strong person is an illusion as well. The strength will come only if your illusions are removed. What do you say?"

"You're confusing me."

"Then you shouldn't have come to a psychologist." I laughed. "You could remain in your comfort zone – with your opinion, but you came to me for my help - to demolish stereotypes because the stereotypes interfere with your intention to build your life the way you want it. Please don't be sad. In order to help you, I must put you through the wringer. Will you allow me?"

"All this is very interesting. Please continue."

"There is a saying, "The good made in a bad faith, is evil." What is your opinion?"

"I would argue that."

"Yes you can I beg your pardon, but it was a simple trap and you easily fell into it. This is also an indication of strength. Being thick headed is not strength. You are ready to argue but with who? I'm not going to argue against sages. In your willingness to argue you are deaf. Forgive me for being straightforward."

"Explain."

"There was a 'saying', but you heard a 'statement'. Do you think there's a difference?"

"I think so."

"Your choice was to challenge the sages. It is brave. The question is: What are you going to get from your willingness to defend your point of view. At the same time this 'saying' could make you think and help you to make a right conclusion. The conclusions are needed to detect the right direction for your strength. Any information may be useful if you are open to it. This way you can quickly find answers to your concerns. So let's see what we have here: Inability to hear is a sign of a weakness, and misunderstanding of the perceived information leads to misconception. A misconception is ignorance, and ignorance is a weakness. Might is always right, but you have a weakness. Your potential power is great but it's constantly vanishing through a huge hole. The precious potential of life is wasted instead of being used for self-growth. Let's say that you're in a stressful situation and someone opposes your interests. Let's say that your opponent was nimble, cynical, and mean-spirited (in terms of public morality). In those circumstances your opponent is stronger and therefore he is right, not you. This caused negative emotions in you such as anger and hatred, but you didn't notice how you started to do evil, so, instead of participating creation, you participated in destruction. After that you want the good from the world but on what basis? You didn't care about being right. You didn't take care of your own strength and you gave unknowing permission to the circumstances to manage your business, so you got the natural result that would follow. Now, you're ready to write off the result to chance, but it's legitimate and it has its reasons. All of that within the laws that you don't accept. So it turns out that everybody wants the good to themselves (those who do the good and those who do the evil), but those who do the good so far in the minority. Do you still consider yourself as a strong and that you do good deeds?"

My companion grimaced as if he bit a piece of lemon and said, "I feel like you crushed and flattened me. I don't know what to say."

"You don't have to answer. Let's create a different image when you aren't flattened, but taken apart. Now, let's reassemble you, but together and in the right way."

"Yes, please. It's uncomfortable for me to be taken apart."

"You need to create a new model of yourself, but without illusions and misconceptions. You must be strong and do good despite all the current difficulties in the world. Create your own internal regulations and rules that you want to follow in order to multiply the good and at the same time to be free from illusions. It's time to become stronger."

"I don't understand at all, please help."

"Do you want consistency?"

"Probably."

"Well, that's one of the illusions. There is no consistency in nature, the world, or the universe. Everything is constantly changing. Therefore, one of the rules - *Have the willingness to change and always be ready for something new.* Are you ready?"

"I think so."

"Would you accept the law - *Avoid grievances, frustration, and judging others?"*

"Yes."

"Would you accept that the ability to listen is more important than talking?"

"Yes."

"Are you writing it down? Great! Another law is - *Accept this world as it is and people in it as they are.* Alright?"

"Got it!"

"Now let's see if we reassembled you correctly. Imagine this: An intersection of life. The good moves ahead but from the side the evil runs a red light. In a moment there is a collision. The evil can die from the contact with the good, but it can also multiply. What allows the evil to beget more evil?"

"Based on your philosophy, the evil will multiply if the good is fake or weak. If the good in contact with evil manifests evil, then the good is fake. It just thinks it's good, but in reality it's an illusion. If the good is real then it will respond with good to the evil or, at least, won't grow the evil. So, if the evil doesn't disappear, then at least it won't multiply. If the good is stronger than the evil, then in their contact the evil will diminish. If the good is strong, it will find a way to disarm the evil. Did I get it right?"

"Bravo Edward, my congratulations!"

"It's too early for congratulations. I understand it in my head, but I don't see how I can implement it in my life. I agree that it's correct, but my personality... I cannot go against my nature if it's not given to your philosophy then what?"

"You are not very positive about your nature. Are you? A pig is not given to speak and it cannot laugh no matter what jokes you tell. Because it is an animal, it has instincts only. Humans are given everything! Achievements come from pure intention. If you have a goal, then obtain the necessary qualities to achieve that goal. Everything is in your hands. It is your choice - either to stay with the illusions or to change your life for the better. What do you choose?"

"I choose to change."

"Let's move on then. In my opinion, the things that happen to you are lessons in order to improve your personality. You have a conflict with people who you work with. You can change jobs of course, but as you know – with the relocation of the numbers, the sum doesn't change. You and I both know that in any business or at any work there are unpleasant elements. Any adverse events can occur at home, work, or on vacation. Unfortunate circumstances can suddenly appear in any situation. These events have a cause, and not what it seems like at first. How do you protect yourself from them?"

"How?"

"*It's impossible to be saved from troubles by changing something around yourself without changing anything about yourself.* It's possible to create within yourself such a change that some unfortunate circumstances won't bother you while others will transform from bad to good. You cannot do this by changing a place only through acquiring skills on how to manage your life by not allowing yourself to experience frustration or fear over suddenly raised obstacles. Also, you cannot allow yourself to judge people for their actions, but allow them to take their own path, the path of their own evolution. Perhaps one day they will realize that they were wrong, perhaps they will not. Do not worry about it. There is no need to take responsibility for their lives. You can only take care of your own life by not allowing yourself to have reactions which you consider unsuitable for yourself. Moreover, after some time you might change your perceptions. Perhaps you will find that other people were right and your judgment of their actions was subjective."

"What am I supposed to do? Let people do bad things and don't stop them?"

"*Do not show the anger, aggression, or resentment.* This is paramount. However, your own interests, of course, must be protected. Also, it depends on the situation either to take a strong action or dismiss the situation. In both cases it is important what your status is. *Are you a victim or a boss?* A victim would have a grievance or feel that things are unjust. Imagine, according to the Creator's plan, *offenders in this world have a legitimate role – to point us towards the inner issues that we have to work on.* To understand and accept this, it is necessary to have inner strength. A weak person would be fixated on the offense and a sense of injustice, so there would be a little chance of them winning."

"Could you please be more specific and explain again who is weak and who is strong?"

"In my opinion, a weak person doesn't see his weakness as causing troubles, or he sees and chooses not to change anything. A strong person is aware of his shortcomings and chooses to develop, transform, or correct his mistakes and grow. A strong person would choose to make an effort to overcome the weakness that hinders him from becoming a successful and joyful person."

"Have you always been strong?"

"That's the point. I had to do this too - travel the way of errors and correct them, and now the movement continues. There was a time when I also had to deal with all kinds of bad behaviours from people. I began to observe how my "enemies" influenced my emotions and feelings. I realized that none of those people tried to make me sad, angry, or hateful, but my imperfections did. I was looking for a way to be calm and cheerful at all times. First, I moved my attention from the people who did bad things to their actions. I avoided judging people and allowed myself to judge their actions only. Then I shifted my focus from the behaviours of others to my own reactions and feelings. I didn't want to struggle with the shortcomings of others, but I made a choice to work with my own bad qualities which were a response to the negative actions of others. I realized that it was necessary to train myself, and I started to use those people as training tools. Now I can conclude with confidence that I made the right choice and my life has changed for the better."

"Hence, there were bad people that you used to judge?"

"Yes, but the time where I used to think and react wrongly are far behind me. I was fortunate to choose the right path. I chose to expand my consciousness instead of fighting with the imperfections of the world and as a result of it I was able to achieve a state of harmony. Second, I learned how to tell the difference between judging people or judging their deeds."

"What's the difference?"

"By judging a person who committed a bad deed, you allow yourself to be mistaken. After all, a nice person can make a bad decision when stressed. What do you think?"

"Probably."

"OK. If you have a habit of judging people because of their imperfections, you delude yourself and it's a direct route to conflicts and loss."

"It's becoming clearer now."

"When you judge the act, not the person, you have more chances not to have negative feelings for them, so you would be freed from the conflicts and you won't destroy yourself. As you know, negative feelings and emotions mainly destroy the person who is experiencing them. *If you're benevolent to your opposition, the wiser you are and the more you contribute to your good fortune, success, and physical health.*"

"My mind is starting to understand this, but how do I accustom myself to it? I'm programmed to be provoked and feel disgust and anger."

"I know the feeling. I had to overcome it myself. I was looking for an answer. The answer has always existed in all ages and all prophets have spoken of it - it is LOVE. *It is a state of love.*"

"Once, during a meditation on a sea shore, I was able to create and experience the pure state of love. It's an amazing experience and a great feeling. In that state of love I remembered all my enemies, but as kind human beings instead. When I thought of them, I didn't even have a slightest negative vibration in my soul. The warmth and tenderness continued to flood my being. The state didn't change when I remembered people who had done something bad to me. At that point I felt love for everything on Earth: the oceans, sky, trees, buildings, cars, myself, and all the people on the Earth including those who'd hurt me before. This feeling allowed me to understand the reason why we shouldn't judge. To judge is unnecessary because our judgment about the reality is very subjective. *The perception of others depends on the state that people experience in that particular moment.* In fact, *only the state is coloring an event, and less the event itself.* So, it makes sense to work on ourselves. It's important for us to create an acceptable system of rules and follow them with all the strength we have. You will see how would change your consciousness and quality of life, how it will increase your inner strength, and how it will revitalize your soul and body. This all will certainly occur when a person starts to grow in a state of love, constantly caring about their growth. When this state fills the whole body, a person will find harmony. *Harmony is a state of a chronic happiness.*"

"So, the key to a harmony is to create a state of love. Well, all the things you've said seem reasonable and I need time to reflect on it, but right now I feel as if you went over my brain with sandpaper." Edward said.

That was where his story ended. "That's the conversation I had with Edward. It felt like I was arm wrestling. First, Edward asked me to wrestle with his weaknesses while he was watching what happened. Then he joined me and had a small victory. He left the office as a different person – wiser and more skilled. I wished him good luck and after that we met a couple of times every other week. As you know, we fail because of mistakes that we make. When we see mistakes we begin to fix them. When we begin to fix them the problems leave. Edward quickly learned the new techniques on how to control his life and after that he occasionally visited me to clarify some details. Later, he just phoned me once in a while with a report of his new success. He realized that his main mistake was to fight the shortcomings of others without taking care of his own. This chronic mistake caused him conflict and endless fighting with himself. Edward made an effort to monitor his thoughts and control his actions. He became more relaxed and life began to change for better. The good old truth is, '*If you cannot change an event, change your attitude towards that event and the event will change'*. By adhering to this principle he began to experience a joy in his life again."

Victor's story shocked me. I saw Edward within myself. I'm looking for something, but I'm still in the status of a "victim" by blaming "someone" for my problems because I separate people and events by "bad" and "good". It's very difficult to control myself and be able to react without judging others.

"I'm still full of ignorance. I'm a student, but I considered myself a mentor as well." Such thoughts were racing in my head.

Victor lit me up like a torch. My desire for inner change became a bonfire. I have clearly realized that I build my life by myself, by transforming my inner thoughts. I distinctly felt a foothold from which I would flip my inner world and shake out all the trash that I still had.

Bravo Victor! Before I had imagined that I would reached his level, and then his words flashed in my mind, "*Don't evaluate.*"

This meeting with Victor gave me a real push to understand what before was foggy. After seeing Victor, I began to experience interesting and unusual events. I wanted to travel, so I went for a trip the next month and that's when it all really started there...

By unforeseen circumstances I ended up on one of the islands in the Atlantic Ocean which was not far away from the coast of Portugal. I went there as a tourist though the island was far away from the tourists routes. I planned to stay there for a week, but I ended up staying there for six months."

CHAPTER 2

My last meeting with Victor had greatly influenced my interests. I had a greater interest in books on psychological, philosophical, esoteric, and art. All those books contained a deep meaning of life and what was beyond its visible boundaries. This literature had awakened in me a thirst for knowledge. I wanted to understand the strange and explain the unexplainable. I desired to go to the unusual and rare places. First of all, I wanted to find a real teacher of wisdom – a sage.

Gradually, all my interests in leisure had shifted towards a search for knowledge. By this time, all aspects of my life had become successful including my business. I achieved a lot and even more than I had expected, but I had not decided on any further plans, so I thought it would be nice to change my daily routine by going somewhere. I got in touch with my good friends from Moscow and Rostov. They were easy going, so we quickly decided when and where we should go to rest from our "debilitating labours". After a while Lenar, Arthur, and I appeared at one of the most beautiful places in Spain – "Costa Del Sol" on the Mediterranean coast.

One evening we were relaxing at a disco club. Lenar got tired of dancing, so he went to the bar and ordered a drink. They were making an extremely nasty cocktail that tasted like cheap moonshine diluted with lemonade, but my friend enjoyed it. Sipping it through a straw, Lenar watched how Arthur and I wildly jumped on the dance floor, significantly contributing to the novelty in the lives of the local young people. Later Lenar noticed a man sitting next to him who did not "fit" the scene of the youthful club. He was an elderly gentleman with a small gray beard smoking a pipe with flavored tobacco. He looked like a pirate from a fairy tale but without the eye patch and parrot. When the man turned around, Lenar looked in his eyes, raised his glass and said, "Your health, sir!"

"Do you speak English?" asked the stranger.

"Not well." answered Lenar.

"Your friends are funny guys." The old man said.

"Yes." Lenar said while struggling with his memory in order to remember more English words. Fortunately, during the conversation the stranger spoke slowly and sometimes used familiar words. This allowed my friend to maintain a conversation even with a superficial knowledge of the language. Lenar quickly got tired of this dialogue and thought that it would be better to eat than to talk. When Arthur and I, sweaty from dancing, came back to the table and joined the conversation, Lenar was happy to introduce us to his new friend. Arthur was fluent in English and I wasn't too bad, so we had a nice chat. In that moment I did not know that my life was about to change drastically once again.

Our new friend's name was Fiop. He was an interesting character and we were excited to get to know him, but he was more interested to know who we were and where we were from. We told him that we were Russian businessmen on vacation.

As soon as he heard that he exclaimed, "From Russia? Wow!" Then he held up his index finger and said, "Have you heard the name Andrey Podgorodny?"

We looked at each other. "No dear, we do not know him. There are so many people living in Russia." Arthur answered.

So Fiop told us the story that long time ago when he lived in the slums of Los Angeles, he was saved from robbers by a man.

"One night I was coming home late from work. I was in a good mood; whistling and thinking about how I would spend the money that I earned, but my thoughts were interrupted by a gang of robbers. I knew that these people would easily kill anybody just to get money for drugs. I stood still. I was so terrified that I couldn't move. They surrounded me, holding out their knives in a threatening manner and demanded money. Then not far away I noticed someone but instead leaving and avoiding being noticed by the bandits, he approached us.

When he saw that I was in trouble, he asked me, "Sir, do you need help?"

"Yes." I replied with a prayer and wondering what he could do all by himself against six stoned thugs armed with knives. The man calmly asked the bandits to let me go and leave.

In response, the bandits started to laugh and pointed their knives at him saying, "We're gonna cut you into pieces if you don't leave."

The brave man replied, "Let the young man go, and I'll give you something more valuable."

The bandits left me alone and got closer to my defender. I was so scared. I had been upset to lose the thirty dollars that I'd earned during a week working in the kitchen of a restaurant, but the man was going to lose his life. The criminals said, "What have you got? Come on. Give it to us or you'll see your own guts!"

The man answered quietly, "I have to give you something very valuable. It's a lesson. Usually, people pay me, but for you I will give it for free." The bandits got furious and attacked my defender. I could run, but I did not want the man be killed because of my money. I wanted to shout, "Stop, don't do it! Take my money, I'll starve for a week, but don't take the life of this man." A spasm of fear squeezed my heart. I struggled to say anything. I was petrified. But then...

Things happened so fast. My saviour crouched in a fighting stance and when the bandits approached him, he turned into a tornado. Before I could blink, with the speed of light he struck them touching each bandit only once. Five seconds. In those five seconds I heard sound of a whistle, clicks of strikes, crunch of broken bones, and short cries. I saw fighter's flying arms and legs. Five seconds - and six motionless bodies were sprawled on the ground with the warrior standing in the fighting stance. I've never seen anything like this before or ever again. He slowly straightened his knees, shook legs, and began to collect the knives scattered on the ground. Then he went to the trash cans and threw them out. After that he came to me and asked, "How are you?"

I was still trembling from the fear. I wasn't able to speak, and I started to cry. I didn't know how I felt anymore - was the gratitude for saving my money or the fact that he was alive and not hurt. I'd just turned seventeen, and the man was about thirty five. He walked me home and said goodbye. After that I never saw him again. Afterwards I couldn't forgive myself for not asking where I could find him. I searched for him on the streets, checking many gyms where he might give karate lessons. It was all to no avail and I only had his name and country - Andrey Podgorodny from Russia.

That encounter influenced the rest of my life. I saw with my own eyes how a good man with a nimble body and a strong spirit could resist evil and stop it. I didn't reach Andrey's level, but I've supported several schools of martial arts which presented a real philosophy and wisdom. I always tried to stick with the right decisions and never missed an opportunity to support those who needed it.

All these years I've had special warmth for this man and his country. When Lenar called me "Sir", I remembered how many years ago my saviour and your compatriot said that the same way you did. Thus, I will share a secret with you."

Fiop told us that relatively close to the coast of Portugal there is a beautiful place, and after hearing about it I believed that surely this was what we needed.

It was an island where a tribe of unusual people lived. They were terse, beautiful, and unusual because of their primitive way of living, but they looked like people who had aristocratic roots. In other words, they didn't resemble the sort of people who might use the simplest tools of the civilized world, but first world people who had rejected all the unnecessary things that were unrelated to the nature of Mother Earth. It could be seen in their manners and appearances. He had never seen anything like it in his whole life.

Fiop was an official person of the administration of Canary Islands, and he explored all the islands within a radius of three thousand miles. On all the islands of Spanish and Portuguese colonies everything was simple and clear - on some islands the Europeans had set up the style of life where everything was "civilized". In the other settlements there were people who received benefits from the government and they accepted the tourists who wanted to experience the natural and simple way of living on the islands. The residents of these settlements were lazy and they used house wares that tourists had brought. There were also islands that were inhabited by savages who were armed with spears and arrows and they stubbornly refused to use the basic facilities because of their conservatism and traditions.

The inhabitants of the mystical island chose to live close to the land, but were different than those with the spears. They had rare abilities and spoke only if it was really necessary. The best of all, it was very hard to reach that island. The nature protected the island from uninvited guests from the water with the reefs all the way around the island and from the air with opulent vegetation. To get on the island it was only possible for those, who knew a special passage and a certain time when it opened. Possibly, a helicopter could land on a narrow strip of sandy coastline, but many attempts by pilots were unsuccessful because of unforeseen circumstances. Though it might seem strange the pilots refused the opportunity to earn money and would find convincing excuses not to go there. There were rumors that nobody could get there without permission of the leader of the island.

Where the inhabitants came from, Fiop did not know. By the appearance of the islanders it would be right to assume that they came from the Spanish or Catalan nobles. The language they used slightly resembled Japanese. The music and singing was similar to Tibetans which Fiop heard a long time ago when he was traveling in China. The rituals that they sometimes performed reminded him of Indians. Their festive dances in some ways looked Mexican. The names they had could be found in any part of the world. It was a beautiful island with extraordinary people.

Such a view of these people had stuck with them when he was there for the first time a few years before. Fiop told us how to get to the place in Portugal on the Atlantic coast, where we could find a man named Kharlampy.

"This man is all you need to get to the island", said Fiop. "As far as I know Kharlampy is the only person who can arrange this event, but I haven't seen him for ten years and I don't know either he's alive or not."

"Is he that old?" We asked.

"No, not in society's opinion, however, there are some interesting things about him. First, he flies on an old leaky "barrel" with wings that optimists would call an airplane. Second, Kharlampy is a native from the island. They don't reproduce very rapidly and usually don't live long. Everything is different with them. As a joke, I call them aliens. I'm telling you that they're very unusual people. Perhaps, you will see them."

"Why are they so unusual? How does it manifest?"

"You will find it out yourself. The main thing is that for those few people who manage to get on the island it improves their lives. I see no reason that you shouldn't try your luck."

It sounded intriguing. My friends and I immediately decided to go the next day. We used a lot of napkins to write the names of places where we were supposed to go until Fiop nodded affirmatively. We thanked him for the interesting information and we shook hands and went outside to get some fresh air.

We were glad that our plans had changed and after we discussed everything we had heard we rushed to the hotel to look at a map and calculate the best route. We found out that the destination was near and the route was easy. A taxi would take us to the bus station where a bus could take us to the port Faro. Not far away from the town should live our potential guide and that was the most important part of our journey because everything depended on the mysterious person. "Now sleep and tomorrow we'll hit the road!" I said.

In the morning when we met at breakfast, my friend's previous adventurism was replaced with common sense.

"What if he made it all up? Anyone could do it just for fun. Maybe some Russians ripped Fiop off and now he sends all Russians to the middle of nowhere to waste their money. Is there really such an island? We don't even know its name", Lenar said reasonably.

"And does Kharlampy really exist?" Echoed Arthur. "It's hard to believe that a place so close to civilization can be something so special that isn't already full of tourists. Can something like that really be a secret these days?"

I didn't say anything. All these reasonable arguments did not affect my decision. I felt an irresistible urge to go. I was not in the hurry to argue and I calmly listen to the arguments of my friends knowing for sure that in any case I would go alone. I knew that the island did exist and I would find the way to get there.

"But what are we gonna lose?" Lenar said almost whispering. "Well, if we don't find the island that's fine, but if we don't go there we'll never know."

"That's right, we should check it out. So, let's go then? How about you Alex?" Arthur asked excitedly.

"Easily", I sighed with relief. We poured some cider in our glasses which was as sparkling and bubbling as champagne. Then we toasted with a neat sound and with a great appetite started to eat looking forward to experiencing an interesting journey which was far more attractive than commonplace entertainment at the hotel.

After breakfast we packed up our bags and called a cab. On the way to the bus station we were discussing the route and the necessary means of transport to our destination. Anything new to us was a joy.

The route was really simple. From the bus station in Marbella we took a direct ride to Faro on a comfortable bus. As soon as we got off the bus, the agents in a colourful headband vied with each other to offer us a taxi. After a short bargain, we made an inexpensive deal with a driver - a young smiling Portuguese man. We had an interesting observation that the further we got from the Spanish border, the closer we got to the beginning of the twentieth century. It seemed as time moved slower there.

By the evening of the same day we arrived in Sagrish where Kharlampy was supposed to live. Our driver stopped the car near the hotel "Baleeira". It was a decent hotel, and we liked it right away, so we chose to stay there without seeing the other two hotels that were in town. As we were prepared to look for Kharlampy for a long time (or it might turn out that we would not find him at all) having a hotel would be a good idea. We really wanted to take a shower, and we were pretty hungry as well. We had no doubt that even if we did not find Kharlampy, we would look around for what we could do there.

After a shower I got dressed and went down to the lobby. My friends were already there, and they were discussing something excitedly. When I approached them, Arthur clapped me on the shoulder and said,

"Alex! You wouldn't believe! We met a clerk named Raul, and I asked him where we can have a decent snack. After he told us several places I thought to ask him, just in case, if he knew a man named Kharlampy. It seemed that he knew him, can you imagine? First, we assumed that maybe it is a different guy, but no, from what we've figured out it might actually be him and he told us where to find him. Fiop didn't deceive us. Kharlampy and his plane are in the fishing village just a few miles away from here. Raul phoned his brother in law to take us there, so he'll be here in an hour. We'll get to the village before the dusk. We should probably leave our luggage here and explore first. What do you think?"

"Sounds good."

"One more thing! Raul has recommended to us a good restaurant called "Last Chance". What a name for a restaurant! This is a sign. Lenar and I would like to go there."

"Me too."

"OK then. Let's hurry. We're getting picked up from there."

The weather was lovely. We set out and ordered a big meal with a fish and some unknown exotic seafood once we got there. With the seafood, we had a very nice white wine. Shortly after that a car pulled over to the restaurant. A young driver got out from an old "Volvo", waved to us in greeting, and gestured that he was waiting for us.

As we approached the village where we would find our guide, we observed a picturesque view of the stretching wide and curved coastline. It was quiet. The sun was preparing to hide for the night at the edge of the ocean and painting the neighborhood in the colour of a ripe persimmon.

To find Kharlampy was not that hard. We saw from a distance a seaplane moored at the pier. Apparently, it was his plane. We were impatiently hurrying the driver. The "Volvo" slowly moved down the slope of the shore and pulled over to the right by the pier.

Fiop was a bit too harsh calling the plane "the old barrel with wings." According to the patches from holes in the body, the battered aircraft with a dragon on a side apparently participated in World War II. However, it was quite a decent airplane.

The mysterious man was fixing the engine on the left wing using a headlamp for illumination. Kharlampy was a skinny man of indeterminate age. He could be forty five or seventy years old, but he did not have a single gray hair. At the first glance we could tell that he was an unusual person. He had dark brown and slightly curly hair framing his dark friendly face. We politely said "hello" and told him about our recent conversation with Fiop. We asked him to take us to the island. Kharlampy took off his headlight and first looked into Lenar's and then Arthur's eyes. I realized that he was trying to figure us out so he would know whether to refuse or accept taking us to the island. After, it was my turn. I thought he was studying me much longer then my friends, but the longer he looked in my eyes the softer his face was becoming. Finally, he smiled and nodded.

"I'm leaving the day after tomorrow from Sagres. If you can be at the south wharf at five o'clock in the morning, I'll take you to the island. You're going to stay there for a week."

Kharlampy said those words looking at my friends as if it was addressed to them only, but not me.

"You're lucky. The day after tomorrow there is a holiday on the island."

Our mysterious aviator gave us some important instructions about what we would need to take with us and what we should leave at the hotel.

On the way back to Sagres we were excitedly discussing how well we managed to accomplish the first part of the journey and how to follow the instructions of our guide. We admitted that Kharlampy was an unusual person. Everything that he told us was very informative without the addition of a single extra word. It was obvious that he was an insightful, tranquil, and benevolent person. It was a great start and we were so excited to get on the island with the strange people and participate in the holiday.

This long-awaited day finally came. The four of us were comfortably sitting in the twin-engine seaplane ready to take off. I was excited. After a few minutes of acceleration our plane pulled away from the water and slowly began to levitate turning towards the open ocean in the direction of the unknown.

From a bird's-eye view we admired the turquoise clarity of the shallow water and the shapes of the shoreline. The sea shore was gradually disappearing and on all sides was the vast ocean. Inside me echoed one word, "Hurry!"

During the flight I fell asleep, and I dreamed of an island with a sandy beach, palm trees, and an exotic and mysterious princess in a bikini made of a sea shells wearing an ancient Russian headdress. She came out to meet us with a bun and jam. There was also Elvis Presley's music in a waltz pace. With the great tunes of the rock star in the background the dark-skinned natives with a daubed faces, feathers, and spears, dancing a waltz while the sand dust rose up from under their feet.

I was awakened by Arthur. The sun was shining, so it made the sea seem to sparkle brightly. In front and below us was a shining green island fringed with a white strip of sand making it not just an island, but a mystery as well. From the distance it looked really tiny. My heart beat with joy – there it was the long-awaited island!

I could still hear the heavy roar of the engines in my ears when we stepped on the logged pier of the island. I felt like burying my head in the sand, so I could get rid of the intrusive tinnitus, but soon the unpleasant sensation was gone. In just a couple of minutes four men emerged who greeted us and went to help Kharlampy to moor the plane and unload the cargo bay. The pilot brought a big load for the islanders.

My first observation of those men was that they are half-savages, but in the way they held their heads, their postures, and their expressions were something as you might see in a Royal court. At first sight it seemed strange and even ridiculous. I looked around. There was no princess, no buns, and no jam.

"My friends, I congratulate you with your arrival," joked Lenar, "Make your mugs cheerful. Don't forget that we came to the party." He added.

As soon as I put my feet on the island, I immediately felt an exceptional comfort in my soul. "Here is a good aura" I thought. I noted that everything around me was touching and at the same time surprising me. I also had a strange feeling that I had been there before. "What does it mean? What kind feeling is this, like I'm returning instead of arriving for the first time?" Such thoughts were floating in my head. The feeling was a little nagging but pleasant, and I did not want to think about it for long. I just relaxed and thought that the explanation would come later. I felt that I already loved the island and loved all the people who were there.

As Kharlampy instructed us, we arrived without our luggage. Clothes, documents, and money we left at the hotel. We had nothing but a minimum of personal care products and some gifts for the leader of the tribe.

Kharlampy took us to the leader. He said that we were supposed to call him "Tare Mugu". Later I realized that "Tare" was something like "the honorable" and "Mugu" was the name of the leader. It seemed like they did not have last names. We went to Mugu's residence - a bamboo hut on logs. The place was very small. The leader did not look like a feral man or a leader of a tribe cut off from others. He sat cross-legged on a low wicker chair which obviously was locally made. The clothes of Tare Mugu were fairly modest and resembled a Punjabi outfit from the East India. He had no rings or earrings, and he was wearing only a set of beads around his neck under his clothing. From the clothes it was impossible to determine that he was a leader, but by looking in his eyes one could see a strong mind and spirit. He was an unusual man. My body responded to the extraordinary sensation of his presence. From this man was coming a quite palpable energy and power, and it was worth my admiration.

Kharlampy invited us to sit down on the floor. We greeted Mugu and then our mouths opened as we started to observe Kharlampy's and Mugu's communication - a silent conversation. I had previously noted Kharlampy's reticence, but the way he exchanged information with the leader shocked and fascinated us because their chat seemed almost mystical. It looked as if they preferred to convey their thoughts instead of using words. I had a feeling that those rare words what they uttered, were more for us so we wouldn't get confused. Later I realized that the word language they used was only so they would not lose their knowledge of oral communication, but they did not really need it. We never saw anything like that in our lives, so we were astonished.

After talking with Tare Mugu, our guide announced to us that we arrived here at the best time because the leader was in a great state of a mind and he was glad to see us. We congratulated the honorable Mugu with the holiday, gave him the gifts, and expressed our desire to stay on the island for a week.

Leader Mugu nicely replied, "Matuar yaka." Kharlampy translated to us that he welcomed us and invited us to stay on the island to participate in the festival.

That was the language of those lovely people - in a couple of words was a long sentence. I have read that in monasteries of the martial arts and spiritual practices of India, Nepal, Tibet, and China, the students learn how to understand a teacher and each other using minimum words by cultivating their mind in a certain way. On the island though, I'd encountered the highest method of a communication - the ability to communicate through thoughts.

Kharlampy sat down next to us and shortly after that three people entered the room - two men and a lady. Her age was difficult to determine. I thought she was eighteen years old but if she was thirty, it would not surprise me. They did not notice our presence and sat down at the feet of the leader.

"Tare Mugu called his close relatives in order to accompany you while you're on the island", explained Kharlampy and added, "Each of you gets a guide." Then Kharlampy leaned close to my ear and quietly explained that during the special holiday everyone who came to the island would be treated with respect, but I was especially lucky because my guide during our stay in this blessed place was going to be a beautiful lady named Ilistre - the niece of the leader. Ilistre, even with her modest appearance was pretty, and the most interesting thing I found that her manners manifested dignity and self-sufficiency not at all typical of indigenous people.

I observed everything that was happening in the room, and the girl especially. For the whole time all three guides didn't take their eyes off Tare Mugu. It was evident of the reverence they had for their leader.

When the discussion was finished, we all went together into the village which began at the residence of the leader. A fairly large village was stretched deep into the island. As we walked through the village, our guides and Kharlampy were exchanging glances with each other - probably discussing something. It was evident that the decencies of this island were different from ours. None of the guides looked at us even once. The men were middle-aged, but the age of the lady I was still not able to determine. Either she was young and extremely wise or she was a mature woman, but well preserved perhaps because of her active life style, fresh air, and healthy food. So, there they were - those extraordinary mulatto people with dark chestnut hair: silent, smart, restrained, and smiling.

None of the residents looked at us when we showed up in their village and no one paid any attention to us as if "white" tourists appeared in their village every day. They were busy preparing for the holiday. While they worked, they hardly talked, but looked frequently in each other's eyes just as Kharlampy and the leader had done earlier.

I had read about sages in the East that could communicate telepathically with each other over any distance, but before me was a whole village. Fiop was not exaggerating by saying, "These people are more than unusual".

The nature fascinated me so strongly that I felt as if I was there alone. I had no desire to discuss anything with my friends despite the fact that they were close to me. When I looked at Lenar and Arthur, I noticed that they were turning their heads side to side marveling at the views and forgetting about everything.

I don't think the vegetation caught our attention. Palm trees, beaches, cacti, and a variety of exotic flowers were perfect, but we have seen them before. The agaves were larger and the leaves of the trees gleamed with a brighter gloss. Could the bright colourful parrots that were walking busily along the ground as like pigeons impress us? No. Flora, fauna, or the exotic places that my friends and I had visited earlier could not make a strong impression on us. Not even the amazing cleanliness of the village could. So what exactly was so unusually attractive about the environment? I could not explain it to myself. Maybe it was the atmosphere, air, or energy of the island that had such an effect on all my senses causing a pleasant feeling. Apparently, my friends were experiencing the same thing.

Most likely, the leader had given instructions to feed us because somewhere nearby we saw variety of delicious food. When we reached the kitchen, in the shade of a palm trees right on the ground the food was already served for us on a large plates made of dense and dry leaves. In the centre was a big bowl with assorted colourful delicacies.

"They have something like a commune." Lenar said.

"Guys, pay attention to how it works, we only just arrived and dinner is ready. If they brought some tequila once in a while, it would be just about perfect. Many cacti are growing here, so probably the natives make moonshine from them." Arthur spoke.

I noticed that Kharlampy responded to Arthur's comment as he understood Russian. He looked at my friend and smiled. He knew that we had some hopes of heaving a tequila party, but he told us to pay attention to other pleasures as having an appropriate meal and not to rush in to scoop up food. On the island every meal was a ritual, and before putting the first piece into the mouth they did a simple ceremony by saying a short mantra. Kharlampy in two words explained and showed us what we were supposed to do. I was fascinated by this tradition. From a simple satisfaction of the hunger and fattening our stomachs, this dinner turned into a pleasant ritual and had a deep meaning of a slow relish, delight, and divine pleasure. I thought of the history of my fatherland. In pre-revolutionary Russia people were always saying a necessarily prayers before eating. "This is a healthy tradition", flashed in my head. After sampling the local exotic cuisine we showed our gratitude to the cook by saying thank you in all the languages we knew, "thank you, merci, danke, gracias, obrigado", but a kind woman tenderly replied, "Ptu-ur".

All three of us turned our heads to Kharlampy - "Please translate."

"She said, 'Relax guys, it is for your good digestion'."

I was ready for further acquainted for the sights, but Kharlampy whispered to me, "Trust Ilistre she is a wise girl. During the holiday we won't see each other because I am going back to the mainland in the morning. See you next week."

When Kharlampy said that, his eyes were smiling. I looked at the islanders with an interest and noted that their faces were calm and relaxed.

I said goodbye to our gracious pilot, and decided to trust Ilistre as he had told me. Kharlampy's recommendation was good enough for me. I was amazed that every time when I heard Ilistre's name, I felt a pleasant burn in my chest.

I looked at my guide, and she turned to me for the very first time. We locked our eyes. The girl flashed at me with her amazing black eyes, and I was literally drowned in them. There was something imperious in her look, but at the same time there was a touch of softness, tenderness, and humor.

Upon return to the house of Tare Mugu I carefully studied and assessed the exotic lady while his relatives listened to his instructions. "She has an ordinary face and a nice body. Not a great beauty, but how confident she is!" Such thoughts ran through my head.

Then I was ashamed of my thoughts. I saw a deep, secret, mysterious, and alluring vast inner world. I saw that in only a few seconds looking into her eyes. The burning increased. It was probably love at a first sight. I felt as if I had an electrical discharge within my body and mind simultaneously.

Ilistre stood up and held out her thin hand to me. Then she didn't just speak, but sang briefly, "Come with me."

"Oh, she can speak English! If it now turns out that she graduated from Harvard University, I would probably lose my speech would have no choice but talk telepathically." I thought.

I saw tenderness and softness in her eyes, and when I heard her voice for the first time, I was amazed how sweet, melodious, and warm it was. In that short word, and even in just the sound, was manifested the femininity itself.

I got up and took her hand. During these body movements, we looked at each other with our eyes fixed. She did not check me out as was common where I grew up – from head to toes, but she studied me through the eyes looking right into my inner world and "reading" my soul. While she was looking, I did not feel any discomfort and I was even pleased. Apparently, she was pleased as well with what she had seen because she smiled. I noticed that the natives rarely smiled as broadly as Hollywood celebrities do, but they usually smiled with their eyes. It was a unique phenomenon: a subtle movement of the facial muscles and it changed the entire face - eyes started to glow with warmth and joy.

"Ilistre", she said and put her palm to her chest.

"Alex", I said and realized that we had just done the introduction ceremony, but we did not welcome the leader and Kharlampy. We forgot somehow and they did not ask. Apparently, they did not care about it, but they could look into the eyes of a stranger and see much more than a person's name could tell.

"Alex". She repeated my name, took my hand, and we walked away from my friends. I felt as if my hand was melting from touching her soft fingers.

My friends called me, and I looked back. Lenar and Arthur laughed and showed me thumbs-up trying to say that I did a good job. They also grimaced silently and made sad faces because they got men as guides. I tried to smile, but I made a serious face instead because it wasn't like that at all.

Ilistre took me away from the village where not too far from the sea was her "bungalow". She pointed at her home. Two meters above the ground I saw a little tree house made of bamboo. The roof was made of dried palm leaves that were tied up together. A house was somehow attached to the mighty tree and looked more like a large birdhouse. We climbed upstairs to this lovely exotic one-room apartment where I saw simple belongings of a simple lifestyle. Some items such as scarves, blankets, pins, and scissors indicated that it was merely simple not primitive. However, it was a quite nice and cozy place. The hostess told me to leave my backpack before she took me to the creek to cool off from the trip.

It was just what I needed. I took off my T-shirt, shorts, sandals, and joyfully swam naked after I smeared myself with special clay that Ilistre gave me. The Islanders added something to the clay and used it as soap. Maybe that is why their skin was smooth, so it was hard to determine their age. I was not embarrassed that my guide was watching me with interest, and it was nice that she was not embarrassed of my nakedness.

After swimming, Ilistre and I went back to the village, but by a different route. She decided to give me a bit of a tour to show me the island. I was in a playful mood, so I happily agreed. A few minutes later my playfulness was gone because in front of me I saw a snake hanging from a tree and gazing straight at me with its forked tongue extended. I froze and looked at my Ilistre with eyes full of fear. She calmly smiled showing me that benevolence is better than fear.

It was my first lesson in nature that contained a deep philosophical meaning. This lesson explained the main principle of building relationships with the entire world. At that moment I realized that it was not just a pleasant walk with a cute aboriginal girl in the woods, but I had the added benefit of delving into the process of studying the world, so I gladly did just that. Ilistre studied me every so often by looking deep into my eyes.

Ilistre did not speak English. She knew a little more than a dozen English words. I tried to understand her by the tone of her voice when she was explaining something to me. Sometimes it worked, and of course hand gestures and body language helped. She was pointing at an object and said its name in her own language, and I repeated trying to remember. So from the first day on the island I began to study the local dialect which was similar to Japanese, but it was little bit more melodious.

We returned to the big meadow in front of the leader's house which was covered with half-trampled grass, and we took part in preparation for the holiday. Women were making garlands from the flowers that grew there in abundance. Men were building something from the bamboo poles. They had axes, machetes, and knives. Skillfully they firmly tied together the bamboo poles with vines. Smaller elements they tied with the coconut ropes, and they did it masterfully. My friends were helping them, and were enjoying themselves much like they were children.

I noticed that women were dressed in colourful skirts and t-shirts, but the males had all kinds of clothes such as faded shirts and shorts, old sailor's uniform, and loin cloths. My guide "handed" me to the men to help them saying only one short word that I did not understand. They warmly welcomed me to their team and I joined the local building trade. I was shown what I had to do while they said unfamiliar words, but I understood one thing for sure - I was more of a hindrance than help. The builders made me think that it was a pleasure for them to enable the savage from the civilized world to take part in this professional construction.

The dusk had arrived. All the natives of the island gathered together. I had not imagined that there were so many of them. The participants were sitting on the ground facing the scene. It was very unusual because there was lack of hum of the human voices. I could only hear a soft rustle of their clothes with the accompaniment of a sweet music of the nature; however the atmosphere was full of solemnity. My friends and I were separated though we could still see each other, but unable to communicate for there was too much space between us. Perhaps it was the intention of the wise leader. Ilistre took me by the hand and we went to the stage where six musicians sat with many drums. The drums were all different sizes - from really big ones to a quite small. On the stage was a throne for the leader. Everybody quickly found their spots and waited for the beginning of the festivity. I sat on the ground with my legs crossed. All heads were turned in one direction – to the residence of the leader where he was supposed to start the holiday. The silence was broken only by chirping of cicadas, sweet singing of the tree-frogs, and unforgettable tapping of the palm leaves. I was enjoying the sounds of the nature. I occasionally glanced at the bamboo house, and more often at my companion. I strongly felt my interest flaring in regards to the mysterious woman. I was attracted to the mystery in her, and I had a wild desire to understand the mystery in her. Ilistre did not notice my attention to her and I felt comfortable when she was close to me, but I also felt like a schoolboy who had his first crush. The analytical part of my brain said I want to hug her, touch her hand, look in her eyes again and again, to hear her sweet voice, but the same time I did not have any lustful impulses. There was a feeling of great respect for her and a willingness to accept her terms of interaction.

My thoughts were interrupted by a wild choral scream from the men who broke the silence so suddenly that the birds flew out of the crowns of the trees chirping loudly. Startled I thought, "That is new. What a surprise - they can shout." They were cheering the leader Mugu who had come out of his residence. Then I experienced a familiar feeling in my body - a vivid sense of power and grace that was coming from the atmosphere, and had touched my essence. My body gratefully responded to that touch by vibrating in every cell.

The scream subsided as suddenly as it had occurred. The leader got on his throne and softly yelled, "Machakeee Ranaka" - the holiday had begun. The torches erupted with a bright light, and the music started to play. There were violins, flutes, memorable drums, and a set of bells which endowed the music with a holiday-like atmosphere of special solemnity.

It was something that I could never imagine even in my most vivid fantasies. I was not particularly a big fan of holidays, crowds, and loud thundering music. I used to quickly get tired of it because my soul wanted something different. What I was experiencing in that moment, was what I had been looking for! It was lifting my spirit with its simplicity, tastefulness, and colourfulness.

First, the islanders sat on the ground, swaying to the music, and clapping their hands in rhythm. Then the bell stopped and had been replaced with iridescent female voices that delighted me with their high vocal performance. At the same time the other group of females started to dance on the patch in front of the stage showing the amazing fluidity of their bodies. It was spectacular! In those perfect movements were belly dancing, tango, and elements of erotic dancing - there was challenge, passion, and fire. I do not know why, but the Spanish ladies with castanets on the coast of Costa del Sol made me bored. But this! In the islanders' dance it was impossible not to feel the admiration because it was filled with so much grace, youth, and femininity!

After a while, there was a loud sound of a bell and the men hit the drums calling to dance. Everybody jumped up to their feet. Ilistre quickly pulled me up by the hand and the entire island came alive. At first, I could not take my eyes off her, and I stood as an electric post because she turned into a dancing hurricane. She was an amazing dancer. She shoved me in the chest and yelled for me to dance and I finally came out of my stupor realizing that I was behaving exactly as a shy schoolboy.

"I have to get involved in the dance." I said to myself and looked around trying to copy the other men. Slowly, I began to dance. I had a desire to learn the local dance, the right movements, and to merge with the music. After a few minutes, I was not looking at the other people and just letting my body move. I was having fun and moved as I wanted with the catchy music. This was not just a dance, but a powerful wave of joy which was expressed through the movements of my body. I clearly felt those waves through me and they caused amazing feelings. The dance had different stages. In the beginning, the dancers moved as if they were shaking something out of their bodies. Later, the movements became faster, more sophisticated, and smooth. Soon, I was adapting to the new movements and then I was immersed in the meditative state. I was within myself. I could feel every cell of my body. My body moved almost by itself as if I was observing my body from the outside. I was able to reach a state when I totally enjoyed myself and I did not need anyone at that moment. Ilistre was out of my sight, but I was not worried. I was comfortable with my feelings of being in love, and I had a pleasant sense of anticipation for the upcoming events.

At one point there was another sound of a bell and for a few seconds it was silent. When the music ceased, everyone stopped dancing and in the silence I heard a strong sound. It resembled a Tibetan bowl. The sound was accompanied with a great instrument similar to a violin. Then I heard a charming and clean voice. Although I could clearly see the leader, I was pulling my neck to make sure. Yes, it was him. I stared at the Tare Mugu trying to get a better look how he sang. Then I realized that it would be better just to listen. I felt someone's touch and I turned around. Ilistre made me a sign trying to say, "Close your eyes and relax." I looked around. Everybody moved with their eyes closed as if they were in time-lapse photography. Some people were gently twisting and some were just rocking from side to side. "Dynamic meditation", flashed in my head. I closed my eyes. I inhaled and exhaled reaching for relaxation. I felt that I could easily concentrate on my body, breath, and the beating of my heart. I felt that all the processes in my body were perfect. I felt the perfection of all my limbs and my whole body in general. I felt myself weightless for I was resonating with the voice and the music. Focusing on the relaxation, I allowed my body to perform its own intricate bends and twists. This thrilling state where I observed how my flesh moved spontaneously without the control of my mind, allowed my body to work on the harmonization of all my muscles, ligaments, joints, and spinal cord. I was surprised to experience a pleasant pain in the muscles, languor, and delight. Time had disappeared. I managed to totally immerse in the condition "here and now" which allowed me to feel the changes in my inner space.

After a short pause, the leader Mugu changed the tempo of the song and I obeyed letting my body go to a state of absolute rest. With my eyes closed I slowly sat on the ground in the posture of Buddha. It was unclear to me with what part of my consciousness I noted that I had no thoughts. It was quite easy to achieve this state of thoughtlessness. I had the opportunity to experience a wonderful feeling which previously was unfamiliar to me. I sensed elation. The singing of the leader and a play of the local musical instruments let the entire island sink into the meditation.

I reveled in a secret and cognitive meditation for the very first time. I have previously read a lot about the benefits of the meditative practices, and I had a burning desire to learn them, but it was incredibly difficult to focus because I could not control my thoughts, and to pacify them was possible for a few seconds only. There, on that wonderful island, I had experienced the power and beauty of meditation. For the first time participating in a holiday like that and not understanding the language of these people, the beautiful new sensations took my breath away. Just one day on the blessed piece of land in the middle of the ocean, the unforgettable holiday and the first meditation opened new horizons for me. I noticed something positively new in my consciousness. I saw outward things as if from the other side, and my inner world became bright, airy, and comfortable. After each of these events, I noted that something was changing within myself. My positive traits that were veiled before had become stronger and brighter. Because the holidays were taking place about every two weeks, I quickly got better at meditation.

Sometimes, I meditated in the evenings on the beach with Ilistre when the ocean was quiet. We sat together on the warm sand and she made a special ritual gesture with her hands that she wanted me to repeat after her. Later, when I could tolerably understand her language, she explained to me that this action allowed us to come out of the physical body and "walk" in the astral plain hand in hand. At least that was what I had understood or wanted to understand. Perhaps, even before Ilistre's explanations, somehow I foreshadowed these "walks" not realizing that it was possible. Thus, I was able to get in touch with the mystical practices.

After about two months on the island, I was able to "hear" Ilistre's thoughts when she "spoke" to me by looking in my eyes. I was overjoyed when I asked her by voice mixing with words whether she "said" anything. Ilistre smiling nodded her head.

* * *

With every day, I felt how my love had become brighter and stronger. Sometimes in the evenings Ilistre went to the village and I stayed alone in the hut. I missed her, and I felt that I was bonded to her. Then I recalled Victor's story. One of the central ideas of the internal regulations was to *avoid bindings.* When people are bonded, they become dependent losing their freedom. A bonded person wants consistency. Desire for consistency is immersing yourself in the illusions. The illusions are doomed to failure. Consequently, the bonds are often followed by claims, conflicts, and pain.

I had to make an effort to state in my mind that she was free to go wherever and whenever she wanted. I stated to myself, "She is a free woman of the free tribe. I can love and enjoy the opportunity to feel this emotion that fate has given me. True love brings joy only, and I aspire to a harmony…"

So, I went to the sea. Comfortably sitting on the sandy shore I closed my eyes and played the music within myself. I spent some time there in seclusion remembering the state which I had experienced during the holiday. I brought my thoughts and feelings back into order that way, and then I was able to enjoy myself – I felt a connection with my soul. Sometimes I was just sitting on the sand listening to the music of the breathtaking lapping of the waves. I could not understand why earlier I hadn't enjoyed the sea as I was then. On the beach, in that moment it was so easy, and I felt so good.

I have visited many different resorts, and I used to come often on a shore at night. I was sitting on the sand or on the bench with my eyes closed trying to turn off my mind and to create a state of calm and thoughtlessness, but it did not work for more than ten or fifteen minutes. I could not stay too long in one position. Everything seemed quite comfortable in those pleasant places, but something had always interrupted me. There was a need to go somewhere or to move. I felt some sort of restlessness as if my ears were plugged, eyes blurred, and brick filled my head. Apparently, something had been disturbing me, but I did think about it long because I did not know what could be different since I would never had such an experience such as what I would discovered on the island, so there was nothing to compare it to. How wonderful that I was able to get rid of the stress! Possibly, the stress had manifested because I was not comfortable with myself or I had a lack of love for myself. There had been a "wall" between me and my soul. It was probably the reason I was attracted to alcohol - to quiet my mind which was unsatisfied with my mundane existence.

On the island everything was different. Maybe because I did not have to rush anywhere, maybe because of the presence of a beloved woman, or maybe due to the presence of something greater.

CHAPTER 3

Despite being a responsible person, I did do something irresponsible. I did not care about all that I have left on the mainland. I lived richly every day and each day was saturated with something new, so it did not occur to me to leave the island in the name of the civilized world. When I thought about home and those who I left for so long, I consoled myself by saying, "I'll go tomorrow or in few days, or maybe next week." I also pondered, "Do I want to take Ilistre with me?" Instead of answering this question though, I asked myself different questions as, "Does she want to go?" and, "Will she be able to live in a strange new world?" My intuition and mind answered together, "No way." Then I asked myself another question, "So what should I do?" and the answer was, "I do not know." I only knew one thing for sure - I did not want to leave her. Maybe later I would figure out what to do, but right then, I enjoyed my life as never before and I was not bored for a single moment.

In the morning, barely awake, I already felt my state of love in the first waking moments immersing in its depths. Every day Ilistre was waking me up with a kiss and each time it was the most beautiful morning. Every night before falling asleep, I planned to wake her up with a kiss first, but I never managed to get ahead of her. Sleeping through the morning I felt a touch of her lips and waves of tenderness and languor ran through my body. In a state of bliss I was slowly opening my eyes as if I was afraid of sinking into the waterfall of the feelings that was flowing down on me from her fathomless eyes. In the mornings a telepathic contact was particularly light and stable. When I was looking into the full tenderness of my love, I could easily read her thoughts, "Wake up my love, the morning is here. A new day has been born - look how beautiful it is." She moved her fingertips on my stomach and chest, and electric discharges occurred in my body. In a matter of a few seconds my relaxed and sleepy state was transformed into the vibrant and vigorous, and we began to fondle each other. Ilistre moved slowly barely touching my sensitive skin giving me an unimaginable bliss. That is how our love games began in the morning.

We had merged into one and I could clearly feel how powerful the sexual energy, the energy of love, was penetrating our bodies and saturating them with incredible power. I lost sense of where my body was, and where hers was. During the love games, we periodically stopped - Ilistre pushed me from her or I moved away, and we devoured each other with our eyes. Like a panther she had fluid movements, licking her lips, and imitating wagging a tail. She was so beautiful! I pounced on her and we were in the ocean of passion again. Our bungalow was shaking, but we did not care if the binding broke and the whole construction crashed down. It was amazing! Perhaps Auguste Rodin himself never dreamed of such a composition where our two resilient bodies had entwined into one with hearts full of love.

* * *

Ilistre taught me a special kind of a sexual-spiritual practice. First, when I got into her, she stopped me and touched my ankle with her fingers. She demanded from me to close my eyes and keep my attention on the touches. She gently moved her hand up my leg and I had to follow the fleeing of tactile sensations. When she reached my pubic area, she switched the touch to her leg and slowly moved her hand down to her ankle without letting me to move. It was very hard to refrain myself and I could not concentrate. I was opening my eyes burning with passion, but Ilistre was adamant. She made a serious face and with her eyes said, "Control yourself!"

While I was getting familiar with this practice, it was difficult to guess where her hand was, but later I felt more and more clearly the sense of the pleasant touch that had shifted from my body to her body. I knew exactly where her fingers were on her body. It awakened in my body even more sensations. This practice developed my ability to feel my partner quite perfectly. After a few "exercises" my mentor and I with a fierceness of passion made love in frenzy. In this sex game I experienced a state of separation from the Earth. The sense of time and space was lost as if nothing existed except for the pleasure that was a cascading flow of feelings and waves of the powerful energy. We had dissolved in each other. I totally felt her and she knew all my desires. We reached orgasm simultaneously, and I thought there was a ten point earthquake on the island.

For a few seconds we did not even move to try and catch our breath. I could "hear" Ilistre's thought, "Why hasn't the hut collapsed yet from our games?" Then we had a wild influx of energy. We got out from our bed, jumped off the bungalow on the ground, and rushed to the beach to dive into the sea where we swam, splashed, and laughed. I tried to race her, but Ilistre was like a dolphin and it was impossible to compete with her. She swam and ducked around me and then surfaced in unexpected spots. After swimming, we raced to the creek to wash the sea salt off in the cool, fresh water. Ilistre was quick and light as a deer and even though sometimes I was able to catch up with her, I knew she was just letting me to do so.

Ilistre had the skills of a stylist. She loved to make me look good. With undisguised pleasure she artfully shaved my beard with a straight razor and after, she would give me a face and head massage. Then, she would gently brush my hair with a sandalwood comb and admire her work. Every time she worked her magic, I would purr like a languid cat unable to contain my pleasure.

After breakfast, we would rest for a little while before going to work in the village where there was always something that needed to be done. In a cleared area away from the trees were small vegetable gardens and banana plantations. The fertile soil of the island generously fed those who lived on it. Often I went with the fishermen out in the ocean to fish. I liked their company, and it seemed as they filled me with confidence and goodwill. After each fishing trip, I came back to the hut truly inspired.

Sometimes, when Kharlampy was on the island, we went to visit the leader Mugu. My air guide translated at these meetings. It was interesting to talk or just to be around the leader. When I saw him for the very first time, I was trying to prepare myself for it. I was sure that the leader would ask me who I was and where I came from, but to my surprise he did not. Later, I had a strong feeling that he already knew everything that he needed to know.

At first I asked a lot of questions, but I never got a direct answer. His manner of responding was discouraging and baffled me. During our first conversation, I was puzzled because to my question, "Where did you come from?" he answered, "Do you like sweets?" I wrinkled my forehead trying to understand why he asked me that. "Well, if he doesn't want to answer, it is a secret", I thought. "I do like sweets", I answered distractedly. "And I do too", responded the leader shaking his head and smiling blissfully. I was annoyed and puzzled by this reply. Later, I caught myself thinking that Tare Mugu was an enigmatic person. I knew that my frustration and discouragement were a property of my philistine mind and awakening intuition. I needed to accept him for who he was and tune in to the person I intended to make contact with. I had to be open to what I still did not understand.

Because I had a desire to understand him through total acceptance and openness, I noticed some differences in my perception. After several meetings with him, I discovered that something had changed in my way of thinking. I stopped judging the behavior of Mugu and many of my questions had disappeared. Later, I started to get the answers from within. Very often those insights arose during or after the evening sessions.

Closer to dusk Ilistre and I went to the shore, and she taught me the breathing exercises that were combined with martial arts. Those exercises were simple, and after them I felt clarity in my mind and lightness in the body. I was constantly amazed by the new possibilities for my body. We used to finish late when the island has covered in the darkness of the night, but we could easily distinguish each other on the white sand under the bright stars and the moon.

Such practices were reminded me of the Chinese people. The only difference was that the Chinese people did exercises early in the morning, but the islanders did them at night. The Chinese strategy was to prepare the body and mind for a day of work, but the islanders' plan was to go to bed purged from their daily work. In this exalted state sleeping was like transfiguration. They did not just wake up refreshed, but renewed and rejuvenated. Maybe that was why their hair was not gray, their skin was smooth, and their eyes looked affable and lucid.

I noted that I did not know and never really wanted to know what was the date, day of the week, or a month it was. It was hard to believe, but I never looked at my watch and eventually I took them off and never thought about them again. I had never been so happy for so long. Usually happiness was fleeting, but on the island everything was different. I was not in a hurry. I knew for sure that I had to be there and this knowledge kept me from worry. I got so much useful, vital, and necessary knowledge from my companion and from others on the island. There was no question of whether or not I wasted my time or if I should return home. I felt that I had transformed physically and spiritually at the same time as being extremely happy.

After six months, I was totally immersed in love. Ilistre knew how to love! She knew how to give, and she enjoyed it. It encouraged me to improve myself and to be more attentive. I learned from Ilistre how to love. I have never thought that I would have to leave her one day. Did she tie me to her? Not really. Ilistre knew how to treat me. I had a freedom, but I had no desire to take advantage of that freedom.

Kharlampy visited the island once a week, and I could always leave the island if I wanted, but I never did so.

Life on the island was quiet and unhurried, but there was no boredom. Every day at sunset all the islanders did breathing exercises and exercises for their bodies and because of these practices, they were amazing agile. Also, they looked young, had good physical health, and a cheerful state of mind. I could feel how my body and spirit were getting stronger as the sense of peace and tranquility became more pronounced and deeper. I realized how lucky I was that I had come to the island. This was clearly a gift from heaven.

We spent plenty of time at various practices as exercises and meditations. The meditation during the holiday every time ended with one of Tare Mugu's songs and the sound of a bell. On the shore it ended with a gentle touch of Ilistres' fingers. During meditation we were in a different time dimension. It was hard to figure out for how long the meditation lasted – it could be for two hours or just five minutes. After we returned to our hut, I tried to figure out how much time we spent meditating, but I quit doing that because it was unimportant.

Sometimes during the walk in the moonlight, I felt a strong passion and at the outbreak of strong sexual impulses I would stop Ilistre by blocking her way. I would open my arms for a hug to snuggle up my lovely lady, but every time before I did so, she guessed my intention and deftly slipped out of my grasp. She dodged and ran away calling me, and when my passion reached its highest point, we would finally approach each other. At first, I thought she acted this way just to tease my appetite, but later I recognized the concept of my mentor. She wanted me to make the internal permutations. I had to control my hormonal urges and through the internal effort I would be able to change my priorities.

I realized that it was very important to learn how totally control my body and my desires. If I would act as a "normal" man, where my desires and body controlled me, then I would be further from harmony and God. If I was further from God, I would be more like an animal. I was amazed that I realized this most important truth not among my own people, but among another.

It turns out that the society of the civilized people, where one of the main virtues is to fill the mind with the encyclopedic knowledge, is blind and stupid. People ignore the possibility of the existence of different knowledge and practices to establish the priorities of one's life values. So that society is going in the wrong direction. This direction is all about how to serve the body exclusively. When people build relationships with each other, the relationships between bodies, they do not try to find the treasure within, so those relationships are a road to nowhere. I realized that *the body is a tool of the soul,* and the body must not be in charge. I felt that this philosophy dominated over regular sex where people seek coitus because they are driven by the passion where the body is the only composer and a conductor.

Caught in my arms my lovely mentor motioned for me to hold my horses for a few seconds and to close my eyes. She put her hand on my chest and gave me the opportunity to focus on the heart centre where the soul resides. So it helped me to transform my animal passion for the flesh of my partner to something new – the uniting our souls. Ilistre taught me to enter her body as to enter a holy place. Combining the highest energy centres in copulation, we achieved incredible bliss. I clearly saw a goddess in my partner, and each time I was more and more convinced how infinite the divine opportunities of humans were.

We had sex everywhere we could: in the sea, on a tree, on the sand, or on a log that lay in a few meters away from our bungalow. Wherever it happened, it was a sacred act. We did not let our passion out before we clearly felt how our souls were embracing. In this flight of our souls we were not ashamed of anything. There were no concerns of someone spying on us. Why would they be? On the island there were no prejudices. Wherever and whenever there was desire, we made love.

Then we went for a dip in the creek. We were a little bit tired but happy and carefree. We climbed up the stairs to our cozy home hugged each other and listened to our hearts beat as we fell asleep.

The next day everything repeated, but I was always in the state of anticipation of the next new day, and it was always a joy.

I vaguely remember the day when Arthur and Lenar left the island. It was a short farewell, "Bye, see you. Tell my family that I am all right and I'll be back soon." My friends thought I was mad when I told them that I would be staying, and there would be no talking me out of it. They hugged me, patted on the shoulder, and wished a nice honeymoon. They knew that I was in love and they clearly realized that it was useless to persuade me to go with them.

I cannot remember when I stopped thinking about my home and my family even though they might be unhappy with my long absence. But everything ends, and my vacation on the island was no exception.

On one of the holidays the leader announced that soon an unprecedented storm would be coming and every inhabitant of the island should take care of their homes and bury in the ground their most valuable property, and especially the boats. I helped to dig deep and big holes in places where Tare Mugu indicated. I dragged boats there, buried them, and rammed the earth. We worked in the usual quiet atmosphere - no fuss, excitement, tension, or fear. Ilistre and I also secured our home more firmly. She guided me, but I still did not understand the engineering design of the structure of our house or how it stayed together.

Just a few minutes before the disaster Ilistre told me, "You're a warrior of Light. Now you're ready. I will help you." At the time I did not pay attention to the importance of her words and treated them as a joke had been made. I was giggling, "Yes, I am a warrior of the savages named 'Red Feather'. Alright you will help me, that's great, but you're helping me already by giving me happiness, the science of life, and the full enjoyment of your company." A day later these words acquired a very different meaning for me, and when

I thought about them, I realized more and more how wise the Creator is.

★ ★ ★

At night a tornado went across the island. That night we did not sleep. Ilistre took me into the woods where we found a small hollow where we hid from the tornado. I cut some palm leaves and padded the floor of our shelter with them. Ilistre asked me to bury a machete and some other sharp utensils to be safe. When we were done we sat down. Ilistre put her hands on my head and looked into my eyes completely serene. She hugged me tightly and I heard her blessed voice, "Everything is going to be alright."

"Of course darling, everything will be fine!" I replied hugging her tightly with tenderness believing that tomorrow after the storm everything would be as before.

First, the wind picked up and as its force intensified it became a hurricane. The rain came down on us with the sea water and sand; it seemed as though the tornado was coming straight at us. Everything was twisting and spinning. I could never have imagined anything like it. It became difficult to breathe and I could not open my eyes. It seemed as a machine gun was shooting the dirt and salt water at us. On my back, legs, and head, sticks and small stones were continuously hitting hard. I endured the pain and I was more worried about Ilistre. I covered her with my body thinking only about protecting her. My illusions that it was just a regular storm quickly dissolved. I clearly felt the deadly power of the tornado. "I must protect my lady" I thought, and at that moment it was the meaning of my life. I was ready to give my life for her without hesitation; my main thought was her survival.

Suddenly, through the hum and whistle of the wind I heard somewhere inside of me the quiet voice of my darling, "I have to go. Let me go." My head felt like it was pierced by a thousand needles.

"No, I won't let you go", my mind protested, head pounding, and my brain burning with the fear of loss. In that moment

I felt my uselessness in my effort to give my life for her. Despite my attempts everything had already been decided and my life was rejected. The cruel choice fell on what was most precious for me - Ilistre. I struggled to open my eyes, but I could not. Being aware of my helplessness I protested and held her as hard as I could, "You will not get her! Angels of death, take me!" Through my turbulent thoughts flashed, "Fool!" This was probably a response from those I spoke to. At that moment I heard the loud snap of splintering wood and I felt something strike my back. I felt pain shot through my whole body, and it seemed as if sparks danced before my eyes. The next moment a mysterious force threw me on my back and Ilistre was on top of me taking all the hard blows that one second ago I bravely endured. Through the pain I tried to turn over and protect her from those attacks, but I could not do anything. In the struggle with the tornado, I tried to breathe, but it was impossible. From the lack of oxygen my body quickly weakened, and I started to lose my consciousness. A powerful and unstoppable force was pulling Ilistre from my grasp. With weakening fingers I tried to hold onto her hand, but she slipped from me. I was pulled upwards and spun as the ominous gale tried to tear me to pieces and with a last ditch effort I attempted to save her, but the darkness swallowed me and I lost consciousness. It was hard to say for how long it lasted. It seemed that the terrible night would never end.

* * *

When I woke up, the dawn was just beginning. It took me a while to remember how to think again. I could not understand where I was and what had happened to me. I was in an area littered with branches, leaves, pieces of wood and sticky, muddy sand. With a great effort I opened my stinging eyes, and I could not see very well. Slowly I tried to move. My whole body was stiff, and I was in terrible pain.

"I am OK" I thought as my mind gradually became clear before one thought struck me like lightning, "Ilistre! Where is she? What had happened to her? I have to search for her!" I began to get out from under the debris, but it was far from easy. My hands and feet did not obey me and any movement was giving me severe pain. While I was climbing out, I became exhausted and dizzy. I had to catch my breath before I continued to move. My body was covered with dry mud and my shorts were in tatters.

"Ilistre, where are you?" I screamed petrified with fear. "She's gone!" Answered a voice from within. What kind of thoughts were those? I had to find her, and I looked around. It was still dark yet and my eyes were full of sand, so I could not see much. I decided to go wash my eyes in the creek and to wash the rest of myself in the sea. I was wondering, "Between the valley where I was hiding with Ilistre and the place where I had ended up was quite the distance. It was more than three hundred feet. That was scary. "How did I survive? The beach was pretty close, and I could have been thrown into the sea and drowned. It seemed that in everything there is a will of the world or maybe God's will?" Absorbed in these thoughts I waded out into the water and made my way deeper before suddenly bursting into tears. I fell on my knees and cried; weeping bitterly and howling at the brightening sky. From inside came a dreaded answer, "Ilistre is gone!"

I yelled loudly, "Oooh sea, why? Tell me, what is the purpose? I don't want this!" The sea hit me in the chest with a light wave as if it was trying to calm me down and wash the dirt from my body and the tears from my face. I sobbed, got up, and with a few quick steps carried myself farther diving into the water with the hope that I would swim to exhaustion and then drown and follow Ilistre.

I swam pretty far from the coast at a fast pace, so my heart was pounding hard. Gradually my mind became calm. I stopped swimming, rolled over on my back, and floated. I started talking to myself, "No more. No matter how hard this is, my life continues. My moment of weakness was behind me. I needed to keep searching for Ilistre. The tornado went through the island and probably caused a lot of damage and problems for my new friends who had become like relatives to me. I must find the others because they might need my help." I said to myself and then I began swimming towards the shore.

★ ★ ★

When I finally joined the islanders, the search for the victims was already in a full swing. I noticed that the rescue teams were well organized. We checked the whole island - and found that most of the village was destroyed. The population of the island had diminished significantly. We did not find too many bodies because the ocean had swallowed most of them. Among them was Ilistre.

No one cried. People looked serious and silently were fixing damaged homes and other things. We dismantled the rubble and dug up the utensils.

While we cleaned the island, we found many things - rusty barrels, pieces of rope, severed legs of giant sea stars, fragments of coral, sea urchins, and other dead marine life. This gloomy view of the island made me think, "I understand why there is dead sea life, but why did tornado bring the trash of the civilization here?" I couldn't figure out the answer to it, and I did not really care. The main question that tormented me was, "What now? What was I going to do?"

I admired the tranquility of the islanders, but I could not be as they were. My heart was bursting with loss and guilt because I did not save Ilistre. I would never see her again. I had lost her. I felt grief and I felt sorry for myself. Something had drastically changed in my life. I felt as if I had come back from a fairy tale.

Above the island everything was as before: the blue sky, bright gentle sun, and small clouds. Under the clouds there was devastation. Uprooted and broken trees had significantly changed the landscape of the island - it was virtually unrecognizable. I tried to find the spot where we had hidden, but I couldn't. All that was left of our hut were memories and a few traces on the trunk of the mighty Beech tree where it had been attached. To get there I literally had to wade through broken trees and trash. When I made my way through the mess, I embraced the tree tightly; a witness of our happiness. That was all what has left of Ilistre. I spent some time there leaning against the trunk of the giant tree and listened to the beating of my heart.

In the evening when it was getting dark and the work had stopped, I went to the shore and called for Ilistre upset that she had left me. I was choking from resentment, and I felt an injustice towards myself. She knew in advance that she would die as she had hinted at before the tornado.

That first night after the disaster I sat on the sand and listened to the rustle of the waves as I hoped to hear the sweet voice of my love. Then I fell asleep.

Early in the morning I was awoken by several small crabs. They pinched me painfully with pincers in places where I had wounds, so I had to leave. That morning I did not enjoy the beauty of the dawn because my wounded body ached from the multiple abrasions and bruises, but the worst was the pain in my soul from the emptiness and injustice of the world.

Two weeks later I left the island where I had experienced true happiness due to my misery from the pain and loss. The Island had captured me for half a year and then it had suddenly rejected me, but I was not as before. My inner state was different. My needs and interests were different. There was a sense of loss and regret about what might have happened. However, I thought that it was God's will, which in the moment of loss is always incomprehensible. Since there was no choice, I had to accept what had happened. The alternative to acceptance - the suffering, I categorically rejected. I concluded that it was the right decision. There was the condition of overcoming regret and later came the understanding that self-pity was selfish. The pity for Ilistre was nothing but a trick of my own mind and my own weakness. *Those who are gone do not need a pity.* I felt a deep awareness of the necessity of these events and changes in my life. I was thankful to Heaven for the island and Ilistre who helped me to discover diversity in the world. She played her role and left quietly. I "saw" her smile, eyes full of tenderness, and a farewell gesture. There was no sadness in her eyes while she was walking away half-turned waving at me with her small hand.

CHAPTER 4

Kharlampy brought me back to Sagres. He was in his repertoire - quiet and friendly. He acted as nothing had happened, and during the entire flight he just said, "Everything is all right." I looked at him and said nothing, but I thought, "What is all right?" As if he was reading my mind added, "It had to happen." His words made me think, and I had a feeling that he knew something. Later that phrase was often popping up in my mind.

Before I left, Kharlampy said, "Come over in the morning, I have something to give you."

"Okay."

First I went to the hotel from which, six months ago, I had gone on a "week-long" trip. Raul met me with a sympathetic look and shook his head seeing my appearance. I was barefoot wearing a loincloth (which was supposed to be shorts). I had scars and wounds as well as an overgrown beard. Apparently, it was a pitiful sight. Without a word he handed to me a key for a room. In the storage room were all my belongings which were immediately brought to my room. All of the important things were there: documents, money, and credit cards.

When I entered the room, I found that it looked just as I had expected, as if I had not lived without sheets, bed, and walls with wallpaper. It was as if I had only left the room for a week and had just returned. The first and most vivid emotion was surprise when I went to the bathroom to take a shower. There was a large mirror and only then I did realize that I had not looked in the mirror for a half a year. I had not cared how I had looked. I had not even thought about it. I looked natural. I accepted myself completely without condition. I looked at my reflection and there was a different person looking back. I talked to myself, "Who is this young but haggard and tired man looking at me from the mirror?" I touched my face. Yes, it was me. I recognized myself, but my eyes! I had some kind of new sight which showed a tranquility, peace, and love, but with a touch of sadness. Something new and beautiful had appeared in my eyes.

Then I recalled how I spent my time on the island and how I had lived before that. I wasted so much time in a futile race for material gain. I remembered the night before the disaster as well as the horrible night itself. I remembered how I could not open my eyes and how it was impossible to breathe during the devastating tornado. I was very lucky that my hands and legs had not been ripped off.

I looked at my watch and I could not remember when I put them on. Probably before the tornado, but I had not worn them for almost half a year. Then it came back to me. Ilistre told me to do so, but why? It was strange that they had not been destroyed.

I remembered the words that my beloved lady had spoken in the few minutes before the wind picked up. She had said, "You are a warrior of Light. Now you're ready. Do not be sad about anything. I'll be with you, I will help you." At the time I had not paid any attention to her words. It seemed that she had known the future and it did not upset her. She had known that there was no death. There was only the appointed time of the transition from one state to another - incorporeal. She had apparently known my future as well and helped me to get ready for something important. She had determined that it was the right time to leave the island. The reason why I had come to the island, a plan of the heaven, was finished. I had to leave the island to continue my journey to the other places of the world. She also intended to help me, but how? Only God knew. It seemed as if she was an angel of the subtle world. I could understand that; but no, I couldn't understand. I refused to understand! Why did she die? She could have just told me that my sweet-study trip was over. That it was time to go home. Why did she choose such a cruel way to send me back? But, would I have listened to her? I didn't know. Why was I complaining? To whom was I submitting my claim? Who was I? Did I understand the affairs of the Creator better than the Creator did? "

I took a deep breath, exhaled, and said, "All right, calm down now." A human does not decide when to "leave". Once it happened that was it, and Kharlampy had said the same. When the "contract" on Earth ends, a person has to "leave". A human mind cannot understand the Divine providence. We had to accept and let it go. That was right, but it was hard. It seems like I had a stone in my chest. Nothing I could do about it and my life would continue. Why did I have those thoughts? That was how spiritually advanced people think. I felt that I had some kind of control over my emotions and thoughts, so I did not allow myself to suffer although the severity in the soul was palpable.

I was talking to myself, "I have to take a shower to wash off the sadness and then shave. If Ilistre had accomplished her mission on Earth, mine was still in progress. If she had completed her soul's task: giving me love, strength, and wisdom, then I still had to do something to leave something good on Earth. What did she mean when she had said 'You're warrior of light'? It would seem that there were tests and events on their way. Maybe the purpose of my soul before leaving was to grow? It wouldn't be that easy, but I thought I was ready for the tests and tasks to come. What tasks? I would know one day."

After a shower I looked just fine, but most importantly that I was smiling, my eyes were bright, and I was happy. I looked deep inside of myself and felt the happiness and the state of love within. I was surprised because in the society where I grew up, it is considered to be that after the death of the close friend or a family member, it was normal to experience grief and sorrow, but I only had a sense of the loss of something permanent. A body and ego accustomed to getting comfort from the presence of a loved one cannot get used to the loss - the source of comfort. They cannot accept this fact as the mind and soul can, although there is an ache in the soul. I had changed, and I found it easier if I *didn't allow myself to suffer.* I must do it because I loved myself. The islanders had showed me how to accept such an event even if it was so terrible.

No doubt that my mind had undergone major changes because I felt a huge potential of the inner freedom; the freedom from being bound to something or someone. The period of my life that I had spent on the island had allowed me to make a breakthrough in expanding my consciousness. I felt love for life, the island, all who lived there, to the whole world, Ilistre, and myself. I felt love that was in the present, but not in the past. Ilistre, by a pre-planned necessity had moved to a different place. I was thinking, "She is free, and she does not owe me anything." Ilistre had shown me a new way to see life, and given a new quality of life as well. Love, affection, and gratitude stayed with me in relation to this amazing woman and to her world which I had only gotten to know only a little. I know that in my turn, I had reciprocated by giving Ilistre love, understanding, and a joy of exchanging feelings. I stated, "I do not have to worry whether she was happy with me on the island or if she is happy now. I am sure she is. There is no reason to regret something in the present when in the past I had been giving everything from the heart knowing that I would not be blamed for the tenderness and care. Ilistre is still alive. Now her beautiful and joyful soul walks in the infinity of the universe. She is free and I am free. She gave me what she could and let me go, and I have to let her go as well." I breathed a sigh of relief from these bright ideas and completely exhaled my last piece of sorrow. It felt good! My body was flowing in the grace and bliss. I was in a state of love.

When I was going through my traveling bag, I was curious about my stuff, and I looked at them with an interest. It was hard to get used to my clothes. Especially since I hadn't worn shoes in a while and my feet felt strange in them.

I was wondering, "What now? I have to get back to my regular life. I have to go back home because my relatives are missing me."

It was interesting that I did not care much about it. It is my life, and I go wherever I want to. Then I remembered, "How is my business? I am probably bankrupt because I had been away for so long." Again, I did not care about it. I felt rich. I was spiritually rich which couldn't be compared to. I had total freedom, independence, and the feeling of being able to solve any problem. In such a mood I could conquer the highest mountain.

In the morning I left the hotel with a traveling bag on my shoulder. I was going home, but I had a feeling that I would not be home soon. Before leaving the town, I went to see Kharlampy. I walked towards the sea where his plane gently swayed on the waves by the pier.

"Good morning Kharlampy!"

"Hi", he quietly answered smiling with his eyes before handing me a paper.

"What is it?"

I took a small piece of paper from a notebook all covered with English writings and the Arabic scrawled at the bottom. There were some names and the address. «Tunis, Beiserta» - the first two words that I saw in the text.

"This is the route and the address. There you can find Tare Dianand. He is Tare Mugu's brother."

"Why?" I asked.

"Tare Dianand is Ilistre's father. She asked me to give you his coordinates.

"What?!" - My head pierced with thousands fine needles. "How?! When did she ask?! Why?!" I peppered Kharlampy with questions, and again my whole body broke out with fire.

"Tare Dianand knows about you. If you want, you can go see him."

"Please explain! I don't understand. When did you talk to him and how does a person living in Tunisia receive information from an island where there is no phone? Ah, yes, I get it. You need no phone to communicate with each other."

"It's not important, but now you have an opportunity. You have a route, so if you decide to go, let me know. See you soon."

"See you."

The conversation was over. Kharlampy thought that I got enough information and went into the seaplane's cockpit with a wrench in his hand. I stood still discouraged by this conversation.

They talked to each other without a phone. They were serene and friendly. They rejoiced with restraint, and woe is unknown to them. Somebody from the civilized world would be more upset if they lost a key for their apartment, but on the island the tornado took so many lives, and the islanders just frowned slightly as if they had been unsuccessful at fishing. Frankly, I was surprised on my own reaction to the incident by saying, "My lovely Ilistre passed away. She was not from 'our' world, but I spent six months with her in a paradise. And now, in just two weeks I overcame the loss analyzing what had happened, and I almost didn't feel the sense of loss. I still had some, but the tenderness and gratitude were more dominating. I had the feeling that the people of the island took everything in life in stride. They are never surprised or upset, and I was almost like them. Yes, I had become one of them, and I liked it. I just hadn't learned telepathy yet. It was only possible with Ilistre, and not always. Yes, life on the island had a tremendous impact on me."

I walked away thinking about the islanders, "Why had all the people lived together in a village while Ilistre had lived farther away? The village was in the middle of the island on a hill, but Ilistre's bungalow was close to the sea. Why hadn't Tare Mugu told her to move away from the coast? Why did these questions come to me when I had already left the island? Why hadn't all those telepaths foresee the tragedy? The leader had mentioned the approaching tornado, but he hadn't done anything to save Ilistre. Or perhaps he knew what was coming and thought that it couldn't be prevented? Was it humility or was it acceptance that the events were inevitable? Understanding the future as heavens' will?"

That was something to think about. They had known something that was unknown to us - civilized people with only five senses who had cell phones, microwaves, who had learned how to conquer the depths of the oceans and space, but had lost touch with nature.

I was pondering, "Where am I going now, home or Africa? What do I want? I have been away from home for so long, so it is time to go back. What does my heart tell me? Tunisia! I have tried to find counter-arguments, but there were none. In this case there was no option but to go to Ilistre's father. Yes, I have decided to go!"

Then I felt affection for her father as a friend or beloved family member. The warmth spread throughout my body and warmth in my heart.

So, if the heart and the soul suggested something, then we should listen. I had decided to go and just phone my relatives to inform them that I was on my way home. The last counterargument – I should count my assets. If there was not enough, then I couldn't go, it was that simple. I checked my bank account, and I figured that it would be enough if I was economic. When the last "no" was dropped, I felt a bright and irresistible urge to see the father of my beloved. That burning desire that I felt was, as I understood, a telepathic call from him.

I started to think of the best way to get there, "I could go from Portiman by boat or from Faro by plane." Suddenly, I realized that I had completely lost touch with reality because my visa had expired, and I would not able to go through the passport control in Tunisia.

I talked to myself "What are you going to do? Maybe it is a hint that I shouldn't go and maybe the obstacle is a warning? No, it is my mind says so, but the soul asks and heart tells me to go. *The obstacle is a normal stage to gain the skill of how to overcome difficulties.* There has got to be a way. I have to talk to Kharlampy."

I hurried back to the pier. Kharlampy was expecting me. When I approached him and started to ask, he had already handed me a note.

He said, "Two miles away to the east, you will see a pier where you will find the yacht "Oklahoma". The captain is my old friend. Give him this note. He's headed to Syracuse with a stop at the port Bizerte. Hurry up, the boat casts off in twenty minutes."

"That's great!" I said happily, but in a few moments my joy of a quickly solved problem was replaced with perturbation.

"Kharlampy, why didn't you say that before?"

"Why? This is an idle question, but we can discuss it if it is more important for you, than catching a boat."

I tried to calculate, "Two kilometers cross-country with a heavy bag on my shoulder in twenty minutes. Would it be possible? Now it is less than twenty minutes. I might be late."

"Thank you Kharlampy, goodbye."

"Goodbye." Kharlampy's eyes were smiling.

While I hurried to catch the desired yacht, I talked to myself, "Don't panic, and move quickly but calmly." It was difficult because different emotions and thoughts ran through my mind, "Only a few moments ago I was at a dead end and now my problem is solved. What's that? Who gave me this opportunity? Did Kharlampy know that I won't be able to fly? I didn't pay attention to his words when he said, 'See you soon' because he knew somehow that I would be back. What if I didn't think about it in advance and at the airport I would have had trouble? I wouldn't have gotten into Tunisia! Well, why didn't Kharlampy tell me to go see his friend right away since he knew that the boat was heading to Tunisia? How so? Why did he mock me? And this heavy bag on my shoulder doesn't let me move quickly enough. How sweet it would be to walk there without luggage! Why do I need so much stuff? How? What? Why? - How many questions and claims do I have? I have to quit complaining and stop grumbling. I must calm down. It was my choice go to Tunisia. Kharlampy didn't decide for me. Well, that's alright. Only I have to decide where to go."

I saw a boat from a distance. There was only one boat at the dock shining with its whiteness. When I approached the "Oklahoma", a sailor came out and started untying the mooring lines. I saw a man in a captain's cap and I greeted him as I handed him Kharlampy's note. He slowly read it and said, "I'm glad to do a favour for my friend Kharlampy. Since he asked I'll take you to Bizerte. You will have to pay only for food which is five dollars a day. It should take three days if the weather conditions are favourable. Are you OK with that?"

"Yes captain!" I confidently answered thanking the heavens and Kharlampy.

"Then get aboard."

As a brave sailor, I clicked my heels and saluted.

★ ★ ★

We arrived in Bizerte in the early morning. When I stepped on the long-awaited expanse of a land, it was floating under my feet for some time. I joyfully inhaled the air that the dear person of mine also breathed. I had filial feelings for him and I was happy.

The streets were filled with merchants who offered their wares to the people who were on their way to work. At the open market on the counters the merchants were in a full swing selling a variety of the fresh fruits, vegetables, and hot and fragrant baked bread. From the minarets echoed the call for an early morning prayer. Everything seemed pretty to me in that morning - even the unusual noise and jostling.

After a long stay in the calm and peaceful atmosphere, the bustle of the crowded port initially stunned me. I was a little confused, and I lost the vigilance. At the bus station as soon as I turned away from the bench where I put my bag, it had disappeared. When I discovered that it was missing, I did not realize right away that my bag was gone forever - it was stolen. I was hoping that someone would bring it back and apologize because they had taken it by a mistake. I was so naive. Finally, I remembered that I had returned to the real world with all its inalienable parts and regularities, and it was important not to "sleep". Since it was impossible to find my bag, I decided to find something positive in this trouble, and I quickly did. I gladly found that it was a pleasure to walk without a load. I remembered how in Sagres I was in a hurry to get on the "Oklahoma", and I had had a strong desire to be without a bag, and it happened. So, my intentions began to implement. *Think before thinking*, I remembered the advice of the sages, "It is necessary to keep track of my thoughts. What thought do I have now? It is easier and more convenient to travel. All the important things are with me – a flyer route, wallet, and credit cards."

I missed my stuff, but I felt so relieved as well, like someone had taken, along with a bag, my sins, problems, and pain that had manifested during the cruise to Tunisia. "That is right! I can create in my mind such a program that if someone steals my stuff, that person takes with it some of my problems. So be it!" In fact, I felt even more comfort and ease. It worked!

I crossed Tunisia from north to south without any trouble. Pretty soon I had reached my destination I had arrived in the town of Ben Gardan when the sun gave me a hint that I needed to hurry to find the man named Sherkan before the night.

In Bizerte I could ask a passerby in English and get at least some information, but in Ben Gardan there was no such luck. I knew only three words in Arabic, and it was definitely not enough. I faced great difficulties to find the address I needed. Twice I took a taxi and twice I was brought to the opposite side of the city where I could not find Sherkan. It turned out that the address written in Arabic on a scrap of a paper was not enough for the success of the seemingly straightforward undertaking. It was a shame, but because of the language barrier, I had an unscheduled excursion around the town. I expected to find Sherkan first, and then take care of finding a place to sleep, but in two hours I made a decent mileage in a taxi and on my feet as well. It was useless. It seemed there was no place in this small town where I did not go. I was close to despair.

When I had reached Bizerte, I had cared about one thing only – how to quickly and cheaply get to Ben Gardane. When the bus rushed me to the coveted town making a rattle nose, I flew to a dream. I was looking forward to see Dianand. I dreamed that he would be sitting in the front of the city's gate waiting for me. My tendency to transport into the future creating illusions and dissolving in them, God did not encourage.

The sun already touched the horizon when I decided to take a taxi for the last time that evening. When I showed to the driver the address written in Arabic, he as the previous two drivers knowingly nodded. He brought me to a distant area, stopped the car, and pointed to an alley. It was impossible to drive there because between the houses was a ditch. The driver demanded I pay him. Assuming that I would not have luck of finding a Taxi again, I thought that it I had done enough searching for the day and it was time to think about where I would spend the night and then continue searching in the morning. I told the taxi driver, "Please wait. I'll be right back. I'll just check the address. I need a ride to a hotel."

He just said, "Money". It was obvious that it was his all vocabulary of the English words. I paid him and walked into the alley. I was looking up hoping to see the desired letters and numbers on the house that were indicated in the leaflet. When I jumped over the ditch, I landed on the rocks and barely caught my balance. I found the problem - on my right shoe the heel had fallen off. That was a bad luck. I picked up the detached part and began to think about how this could happen to a low and wide shoe, "How could it all tear off at once? Strange, perhaps, this Italian footwear was made somewhere in Asia. I should do something about it because I won't get far without a heel." I was perplexed thinking what to do when I noticed a stranger - a skinny old Arab. When the old man approached me, he harshly and abruptly said something in Arabic and gestured for me to follow him pointing at my heel. I did not really understand what he wanted from me, but I sensed that I needed to follow him. I hobbled after him holding the hapless heel in my hand. We walked less than a three hundred feet and around the corner I saw a shoemaker who sat on the street putting away his tools - his day of work was over. I sighed with relief – at least there I had some luck. The old man was glad that he was able to help me and held out his hand rubbing his thumb on the index. I gave a dollar to the old man. He was pleased, said something, and left. "Probably he wished me a success in my searches. Why did I think that? Is it fantasy or intuition? No, Telepathy!" I speculated laughing at my thoughts.

Before, I would probably be upset because I wasted so much time for nothing, and especially in the evening when I had to find a place to sleep. Instead of checking out another address, I should settle down for the night. In addition to the unsolved problem I acquired another problem and I was forced to go in the opposite direction. Then I learned one of the "golden" rules, "Why do the small troubles make me worry? What would not seem insignificant in comparison with the course of the river of eternity?" I was pleased to note that I accepted that situation, and that I was calm.

"Hello mister." The cobbler greeted me.

"Do you speak in English?!" I was delighted, and I almost burst with joy and surprise.

"I studied almost a year in Oxford. I wanted to be a lawyer, but I was expelled for a drunken brawl. Give me your shoe."

I gladly gave him my shoe hoping that he would help me with the address I was looking for. I was not in a hurry with my questions. I wanted to hear the former student talk.

"Do you like to party?"

"I like to drink. That's why I'm a shoemaker and not a lawyer. What brought you to this slum?"

"I'm looking for a man named Sherkan."

"If you're looking for something, then you'll find."

"Unfortunately, I don't speak Arabic and that's why I wandered around all day without any success."

"Well, it's not true."

"You're right. If I've met a man here who speaks English, and it is a success. Can you help me with this address?"

"Who gave it to you?"

"A friend, but you wouldn't know him anyway."

"I would like to know the person who gave my address to a foreigner and who walks so terrible that the heel of their shoe broke off."

I did not get it right away.

"No way! Are you Sherkan? But ..."

"In Ben Gardan there are no people with this name, and probably in a whole Tunisia as well. In India, it is different. So, who gave you my address?"

"Kharlampy."

"That's what I thought. I knew right away that you're not an ordinary man. Glad to meet you. I'm Sherkan."

"Alex."

"Now I will fix your shoe and we'll go to see the honorable Dianand."

"I don't think I told you that I'm looking for Dianand!" I said that without hiding my surprise and watched with interest to see how events developed.

"If you got my address from Kharlampy, there would be only one reason."

"Hmm. How mysterious." I muttered to myself. Gradually, my earlier trepidation was replaced with comfort. Everything was going well.

"Please tell me dear Sherkan, what did the old man say before he left?"

"He wished you a good luck in your searches."

"Yes-s-s!" I blurted out. "Great! It works!" I did not know how, but I felt that I had surely learned some skills which were unknown to me yet.

When my shoe was fixed, the shoemaker and I took a taxi and headed out of town. On the way to see Dianand I went over everything that had happened to me, "If I hadn't broken my shoe and met the old man, I would not have found Sherkan because my knowledge of the Arabic language was poor. The broken heel and the old man helped me to appear at the right time and in the right place. Was it a coincidence? What was the probability, without knowing a language, of finding the right person who I'd never met in the wrong place? And how had I gotten to the island? The decision I made to go for a vacation to the place where I met Fiop who actually gave me a "ticket" to the island, and then the heel and the old man - all were links in a chain. Had such things happened to me previously or had I just not noticed? One thing was true - I had become more observant, and that was great."

The taxi turned off the highway on to a gravel road and we were soon approaching a three-story building with a massive fence. The building was surrounded with slender cypress trees and looked like a temple or monastery; it was a building with complicated architecture. Sherkan said something to the driver, and we stopped at the front gate. I paid the taxi driver, and we got out of the car. As we approached the gate, Sherkan looked at me and nodded smiling before winked with both eyes and ringing the bell. Right away a lantern turned on above the gate and a young man who looked Arab or East Indian opened it with a questioningly look.

"Burku Lame", Sherkan said. The young man bowed and opened the gate wide inviting us in. When we came in, I felt warmth in my heart and I smelled a familiar scent. I never thought of what the air on the island smelled like, but I clearly remembered it. Yes, here was the same fragrance as in both Tare Mugu's place and Ilistre's hut. Every day she used to light incense. These incense were one of the export items of the islanders. I inhaled the air with pleasure, and again I felt love for the island. It was precious to me. I felt a strong and a bright presence just as the first time I met Tare Mugu. I felt power and grace and my heart was fluttered with joy.

Sherkan whispered to me, "This is Karim. He will help you. First, you should take a shower, have a snack, and then we'll go see venerable Dianand. Talk to you soon."

CHAPTER 5

"I've expected you." He studied me as he spoke.

"Yes, that's right. Ilistre is right" spoke to himself in a whisper Dianand. I patiently and silently waited. I had a slight shock as I realized that he talks about her in the present. What was she right about? What was next?

"Is it important to you?"

"What?"

"You want to know what happens to you, what happened on the island, and why you came here."

"Yes. You read my mind." I said without a surprise.

"You're quite open. You have a large and bright soul. It's easy and pleasant to read it, but not all that simple. Even in the world of an enlightened soul there might be some rubbish, and the mind is the cause. The aspiring soul wants to get rid of the power of the mind and purify its world. That's why you went to India and that's why on the way there you gained some knowledge through meeting people. For this reason fate brought you to the island where you met Ilistre, and for the same reason you have come here."

"You know about India as well... I'm not surprised by anything but still, it's amazing. Please tell me, the tragedy that happened to Ilistre and the islanders, was it inevitable?

"For how long did you want your peaceful life on the island to last?"

"I enjoyed every moment of that life, and I never thought about it."

"You were gaining the necessary experience for yourself to be able to fulfill the task of your soul. A tragedy – it's your point of view, but in fact, there was no tragedy."

"The fact that people have died, is not a tragedy?"

"Who do you see in those people? - Bodies that need favourable conditions and events? It's a common opinion which isn't helping you to understand and accept God's will and the meaning of life. *A person - primarily is a soul, but not a body.* Every soul has a contract. The contract of those people ended. They "left" the Earth and transferred to a different place. In other words, they moved to another island. Was it a tragedy? No, but there is a *personal tragedy of the soul when a soul outlives the limit of its contract on Earth, and has not fulfilled its task.*"

"There you go. Exactly what I've thought. What is this? Telepathy or intuition? That's interesting." I spoke to myself before putting the question to him. "What task then?"

"I won't give you a direct answer. There is no use in direct answer. The most valuable answers are those that you find by yourself. Everyone can do it. You learn to find them by listening the voice of your heart - the voice of your intuition. You have some success already. Now, I want to tell you something important. The script of your destiny you rewrite hourly by yourself. That's why many of the troubles that are in your fate you can avoid. *With your thoughts and actions you can attract or repel events which might be pleasant or unpleasant.* This can happen consciously or not, but there are inevitable events which cannot be avoided."

"So, there is a destiny after all?"

"The same event can bring a person a grief or lesson, disappointment or discovery, and a priceless acquisition."

"What does it depend on?"

"It depends on the way you approach an event."

"It depends on the way you approach an event," echoed in my head. "Yes, I begin to understand. My fate I make by myself. What I do, how I think, and on what I spend my time and energy for - that's what I get. Therefore, some pleasant and unpleasant events happen as a natural result of my thoughts, emotions, and actions. I get what I desire. In what state am I? Am I prepared? How secure am I? My shield and weapons are knowledge, skills, goodwill, and peace. My weaknesses are pride, envy, hatred, aggression, and delusions. I wonder if my weaknesses have lessened and ideally if they might disappear. There was a time when my body was filled with weaknesses. This result depends on the state that I am in when I meet the event. The more weaknesses, the less chances of getting a positive outcome. Hence, in my nature there is an ability to learn how to manage my life. Please tell me sir, were they my own thoughts, or did I hear yours?"

"Now you're close to awakening. That's why you find the answers. All the answers are within you. *The answer doesn't go to a person who doesn't ask or is deaf and unreceptive to it.* For this reason, until you reach a certain point in your life you don't receive them. When you are willing to listen, you start to seek for the answers and at the same time you start to maintain the purity of your thoughts. As a natural result you begin to receive the answers through the people you meet and now you have reached the level of consciousness where you can find the answers within yourself. That's exactly what you just did. Continue to grow and you will discover a lot more. Would you like to ask me something?"

"Where did the islanders come from? It seems like a miracle."

"I know that you revere a great East Indian saint Sri Sathya Sai Baba. I'll remind you of one of his accurate sayings, *'Do not believe in this world as a reality.'* You are at the service of the forces of good. You were born with it. You came to this world voluntarily. Therefore, it's not important to you what's going on in the outside world – it's an illusion, a play. The important thing is – what's in your inner world. No one but you creates this world. You're the only one in this world - a single all-powerful creator. Take good care of it. Take care of its purity and its size. In this, a man is like to God. Tomorrow you will leave Ben Gardan. If you will *continue to search for an exceptional way to manage your life*, you will multiply and you will never lose the knowledge and the skills that you have earned, but you have to prevent manifesting your pride and those illusions with which you are familiar. There is a common phenomenon when the power of a man or his talent through his success leads to a disaster. This is a possible scenario for you. So, I've just answered your main question, 'Why you are here.' *You came here just for this warning!* Now you're armed. You just have to remember, *'Do not allow yourself the state of triumph, arrogance, or even indulgence over those who are defeated.'* You have to strive to show compassion and generosity. The vertex in your motion – is to keep the state of love.

The state of triumph might become an illusion which causes blindness. By being blinded you may not see that you're in danger. You might not notice how the internal enemy can arise and open the gates of your castle letting the enemy enter from the outside. You cannot also exclude that in this case one serious failure is enough to lose. Danger is the other side of success. The higher you go, the greater the responsibility. Remember this, son."

"Will I always have to win?"

"I'm glad you asked me about it. Even though you know the answer, let me remind you that *everything is in your hands.* If happens that you lose, *do not allow yourself to feel anger and hatred to those who beat you.* Create a willingness to keep the same state and the attitude towards the enemy and to yourself whether you win or lose. If you don't need to be defeated, it will not happen. *The defeat may be necessary to help a person to get rid of weakness, destroy the illusion, and to expand the consciousness.* You need to come to such a state of consciousness when the concept of 'enemy' completely disappears. It is a state of love. It won't be easy for you. Think about it. Now you need to rest. In the morning before you leave we can chat some more."

The powerful energy of love began to fill my mind and the body. It was perfect. It was a state of happiness. I felt as I was filled with love. It was an unforgettable state of inspiration. I wanted to say, "I understood, my teacher", but I said nothing. In the response to my impulse Dianand just slightly nodded and answered with his eyes, "I believe in you, son."

In the morning Dianand asked me with a smile, "Do you want to go home?"

"Yes, I do very much."

"Do you trust me?"

"More than myself, sir."

"It's too early yet for you to go home."

"Why?"

"You're still full of illusions. I see that you have a strong desire to save the world. Am I right?"

"Oh, yes! Great desire! It's the only thing I think about."

"You need to forget this obsession. The world doesn't need to be saved. Perhaps, from yourself only. The world is beautiful the way it is regardless that you disagree with it. The world is beautiful no matter what is in your inner world, but if you would improve yourself the world would be thankful to you. You can count on it. Before you return to your ordinary world - a world of vanity where a lot of ignorance, you need to do something else. You have to become like a lotus. Lotus doesn't get dirty because the dirt doesn't stick to it, no matter where it's growing. Therefore, you need to gain the ability to repel the dirt which surrounds you, so in any environment you will stay clean. You have to strive to achieve such a quality in your personality where you will spread purity and beauty. This radiance is able to clear the space around you. You must regain this ability."

"Is that a contradiction - beautiful world and the dirt in it?"

"A contradiction is what your mind has found based on your current level of the consciousness. It's natural. When your mind with its laziness to think is facing non-standard information, it prefers to reject the new information. Don't be lazy and ask your mind to work hard. Place the source data in the correct order, and instead of a contradiction you'll see the formula. Do it by yourself. Contradiction is a bridge to the truth. It's a bridge to the expansion of the consciousness."

"Is it a task of my soul to become as a lotus?"

"You can awake the sleeping souls and you can light the fire in the hearts of others, but now you're still weak. Spreading goodness by being weak, you are rendering an unkind service. If you're weak, you will certainly become fodder for the dark forces. You must be trained. It will take some time. Are you ready?"

"I am ready", I replied without any hesitation.

"You have to go to Sinai Peninsula in Egypt, and you need to get in the Monastery of Saint Catherine. There you will find an old monk named Mohammed. He will be your tutor. You will have to perform all the tasks of the abbot and the monk.

"Do you remember the state of your soul that you had on the island?"

"You bet! It's impossible to forget!

"Great. There you'll have a completely different life from what you had on the island. Your task will be to reach a state of happiness and love in this monastery where the conditions are not so comfortable. You'll have to work hard, but not with your mind, to understand how the monastery monks can reach such a level of consciousness that after the death of their physical bodies, their bones, which were exhumed three years after the burial, can radiate goodness, goodness which they nurtured and carried in their hearts during their life time. Also, you will have to comprehend why many of them die being young. One day you will know when you can leave the monastery. Don't ask them when. Here's the letter for the monk."

I took the letter and bowed at his feet. Then I got down on one knee and bowed my head. He put his warm hand on my head and I felt a gentle heat, power, and grace flow from his hand like a strong current through my inner space. I knelt in that position for a minute and when he removed his hand, I rose. Dianand also stood up, embraced me as if I was his son and said, "You can go now."

I was so touched that I was unable to say anything. I bowed again and left a room without saying a word. I felt that I had been blessed.

Karim walked me to the gate where I said goodbye to him, and left. Sherkan was waiting for me in a taxi. When he saw me, he got out and quickly walked towards me. He took me by the shoulders with his both hands and said, "Let me take a good look at you." Sherkan smiled and carefully looked at my face as if there was something extraordinary. "Let me hug you Alex. You've met a great man. He gave you some of his power. He trusts you. You don't even realize how rich you are now. Then Sherkan hugged me with a force and happily said, "Get in the car, let's go."

We went to the bus station. Sherkan did not ask me questions, and I did not want to say anything. I was in a state of bliss. My guide was content and occasionally looked at me. It was obvious to me that he was in a perfect mood. Then he asked, "Where are you going now?"

"To a monastery in Egypt."

He nodded with understanding.

⋆ ⋆ ⋆

I continued my journey. I was going to Egypt.

I had no thoughts about what was going to happen there. I enjoyed living in the present moment and I knew that no matter where my fate brought me, I would always feel good there. I knew that no matter what events I faced, I would overcome them all. I had absolute trust in Dianand and the knowledge that it was necessary to follow his instructions. Because of it, I had a feeling that my life was going well.

I arrived in Egypt without any problems with only three bus connections. I was right in time to catch a plane at Cairo airport, and it was during an influx of the tourists. Nevertheless, I got the last empty seat on the plane. After a half an hour, the airliner took off rapidly taking me to my destination. Upon my arrival in Sharm El-Sheikh, I asked the employee at the airport the best way to reach the monastery. Right away, he spoke to someone on the radio and I was directed to a bus with a group of German tourists. I was lucky again. I was getting used to everything working out for me. I was starting to comprehend how things could come together so easily - it was my inner state which I would describe as peaceful and serene. I had the knowledge of rightness of the current events. Also, I was confident in a favourable outcome for my actions. That state that had been developed from the several components had allowed me to get support from Heaven.

In four hours the bus brought us to the hotel "Plaza Santa Katerina". The guide kindly explained to me that we would spend one night at the hotel and in the morning we would visit the monastery which was nearby.

In the morning the bus along with the German tourists took me directly to the walls of the monastery. I stepped on the ancient land where prophet Moses used to preach. A monastery-fortress was tucked away among the silent and majestic cliffs. There were just rocks and almost no vegetation. When I looked at the cliffs, they stunned me with their grandeur, and I could feel the presence of God.

Here it was - my new home and a place of further discoveries. It was pretty crowded there. I bought a ticket and went inside with the tourists. I had a rush of mixed feelings. On one hand I could feel the good atmosphere of the monastery, but on the other hand I felt anxious. I had a feeling that the door was just about to slam after me and I would be locked up. Intuitively, I sensed that days of hardship were waiting for me and I had a desire, before it is too late, to go outside - back to my freedom and to run away from the solid walls. I explained this fear as a manifestation of my weakness which was afraid of new and future tests. I had to make a willful effort to resist the temptation to run away, so I added some humor and romanticism to my mood.

I looked around and I saw a young man who was dressed in a monk's clothing. I told him that I was looking for Mohammed. The young man looked at me with interest, nodded, and walked away. Ten minutes later an old monk approached and questioningly looked at me. I realized that it was Mohammed, and I gave him the letter without saying a word. He took the letter, glanced at it, put it away, and with a nod motioned for me to follow him. It looked like they were expecting me.

Mohammed took me through the narrow maze of halls in the monastery and soon led me to a cell - a tiny room with a couch, metal barrel of water, small writing table, and a small window. It seemed as nothing had changed in the monastery for the last fifteen hundred years and only two things were new - the stainless steel barrel and a mug. The table was more than a dozen centuries old because it was unevenly polished by the touches of those who had lived there before. Despite the fact that the room was small, it did not feel cramped and I found a certain comfort in it. "This is your room," said Mohammed in a soft singing voice. On the couch there was a monastic robe. Yes, they were expecting me.

My monastic life had begun. I had to get up every day at four o'clock in the morning, then I had to pray, work in the monastery, and almost daily climb Mount St. Moses on the monks' path with a heavy load on my back. By the second day my romantic mood had disappeared. The euphoria quickly vanished and was replaced with routine - the unusual style of life was peppered with hard work and the brain piercing question, "Why did I need it?" Only my trust of Ilistre's father allowed me to endure it.

For the first few days I felt a painful contrast between my life on the island and the new conditions of the monk's life which at first seemed hellish to me. Mohammed was friendly to me at first, but the day after the introduction to the monks, he drastically changed in regards to me. To say that he treated me in an unfriendly manner wasn't even close. I felt that he used every opportunity to mock me and to find the most difficult and tedious work that he could for me to do. From time to time he lectured me, "You came here smug, gleaming with pride and arrogance. You thought that you had achieved something, but you still continue to demonstrate your ego which firmly stands on your beliefs and concepts. Here you have nothing to achieve and here is life in service only." Or, "You look tired. It's stupid to carry the extra load of your pride up the mountain. You're not just stupid, but also deaf. I asked you to bring only water up the mountain. Go back and bring only water!"

After the inspiration and warmth of Dianand's speech, I was sick of the mockery from the old monk, so I had a great dislike for him. Gritting my teeth I silently endured his insults. Anger festered in me whenever he did it. In order to calm myself down, I tried to take his words as the harping of a senile old man.

Later, I began to realize that the feelings I had towards Mohammed were nothing but an indication of my own weakness. I started saying to myself, "Anything that happens to me is what I need." Sometimes I was able to admit that he was right. He picked on me purposely to show me my weaknesses. Grudgingly, I followed all his requests in silence knowing that once I got to the monastery - I was obliged to go through all the torments in order to increase the quality of my spirit. I clearly began to understand my task. *I had to learn how to be cheerful and happy in these severe conditions by overcoming painful blows to my ego, the whims of my pampered body, and the darkness of my narrow consciousness.*

I often recalled the island, Ilistre, our love, and the pleasure of a carefree life where everything was perfect. Apparently, it was necessary to balance everything, so that is why life at the monastery was different. There was tedious and hard physical work, mockery from the old man, boring prayers, modest meals, and sleeping less than five hours a day. Moreover, I contemplated daily how curious and carefree were the tourists and pilgrims who travelled thousands of miles in order to perform the famous ritual of climbing Moses' mountain to get rid themselves of their sins.

"Even in the world of an enlightened soul there might be some rubbish", I remembered Dianand's words. "Well, I have a great opportunity to clean the world of my soul up and make my soul shine since I'm destined to be here. I had climbed the mountain more than a dozen times already, and not at night as the tourists did, but in the afternoon when the sun was scorching and while I wore the monastic cassock made of a thick fabric plus my heavy load", I murmured to myself.

Anything I did, I tried to do with pleasure knowing that it was a great opportunity for me to reach true purity of life. My body became stronger on the island and it helped me with those burdens. I earned the ability to perceive everything with kindness.

Sometimes, I made three climbs to the top of the mountain in one day. On such a day, I was barely back to the evening prayer. I was ready to collapse on the ground wanting only one thing - to get to the couch in my cell and fall asleep, but I still had to wash my sweaty clothes.

During the evening meal I was looking at the monks and thought, "I understand what they are doing here, but what about me? While I scavenge after the tourists and carry the water up to the mountain bathing in sweat, I basically waste my time. At home my business probably is falling apart or even worse – has already been destroyed." Due to exhaustion I did not care anymore about the purpose of my stay there. I did not want any truth or spirituality. "To hell with it and my pride! Maybe it's not a pride at all, but the instinct of survival. Who knows? Now, I need to sleep and tomorrow I will think about leaving this place because I cannot do this anymore", moaned my tired mind. I wanted to get away from my torments, and I always wanted to sleep. The next morning though, I found myself thinking, "Well, one more day and then I'll see."

From the very first day I had thought about, how long would I stay and when would I leave? I ignored these thoughts as a sign of impatience and I was growing the humility within myself trying to obtain the ability to be happy where I was. "There you'll have a completely different life from what you had on the island. Your task will be to reach a state of happiness and love in the monastery where the conditions are not so comfortable", Dianand had told me. "Yes, that's right. The master of his fate should be able to enjoy the life in all circumstances and not to be afraid of the problems. It's good that I'm learning it here. It's not so enjoyable as it was on the island, but at the same time this is a million times better than in the "Gulag" prison during the Siberian winter. Compared with those conditions the monastery was a paradise, especially since I had come here voluntarily. Everything is relative. If I was a prisoner, it was only of my own narrow consciousness ", I figured.

From time to time I remembered Dianand's words about the work for creating a state of love, and did my best to be tolerant towards Mohammed and to his taunts. It was extremely difficult, but as the time went by I gradually got used to the monastic life not allowing myself to suffer and forcing myself to stay positive. No matter how heavy the work was, it was time to learn.

Dianand was truly wise man! Day after day I imposed to myself to have a serene state, and I succeeded. First, it was episodically - for only a few hours a day. Finally, arrived the day when I was able to stay in that state from dawn till dusk. I continued to practice my patience, and I felt the difference. I discovered that it was much more difficult to work when I was suffering than to work with love and joy. When I tuned myself to love, it was much easier to climb six miles of steep steps made of unleveled stones under the scorching sun. I really was less tired, and I stopped suffering from the lack of sleep as well.

I concluded, "Therefore, everything depends on the state of the mind. Mohammed was right. I was tired because I was caring a heavy load of false opinions about the perceptions of the world. Because I had incorrect thoughts and perceptions of the world, I was constantly tired. That is right!"

Occasionally, I visited the room where the bones of the monks had been laid over the centuries. When I went there for the first time, it was sad. Visual contact with the piles of skulls and bones lying on the ground couldn't cause joy for me at first sight, but after a few seconds my mood and emotions began to change. Despite the fact that the view was unpleasant, I felt an overwhelming presence of love. I clearly felt a good energy from the bones. I had never experienced anything like it. I just let the energy pass right through me. I closed my eyes and enjoyed the moment merging with the divine vibrations. After each visit, this energy and vibrations felt stronger and stronger. During one of these visits I remembered the words of Jesus, "To love those who love you is easy, what is this job? Try to love those who hate you ..."

I followed this recommendation and I noted that my bad attitude towards Mohammed was gone. I did not need any longer to strive to remain in the state of love. I analyzed, "I stopped judging him! I've accepted him without any conditions." Along with that, I felt my inner strength and comfort because of that. I did not attach much importance to it, but it seemed that the old monk began to treat me much better. He allowed me to access the library where I could find the ancient writings of the saints. I grasped the meaning of the incomprehensible Old Slavonic language by my feelings rather than with my mind. I began to feel the good atmosphere of the monastery and to feel a positive energy within its walls which had accumulated for fourteen centuries.

Once I asked Mohammed, "Where did the tradition of digging out the bones of the monks after three years come from? What's the purpose?"

"This is one of your tasks. You have to find the answer by yourself. When you understand it, then you can leave the monastery."

"What if I understand it tomorrow?"

"Are you in a hurry to go home?"

"I try not to think about it. I try to gain from this place as much as possible. I know that it's good for me."

"Great thoughts. Your stay here will end at the same time when my days are over."

"How is that?"

"On the day - when my heart stops, you will leave the monastery."

"What kind deal is that?!" I abhorred. Then I thought, "On one hand, I have been here for more than two months and of course I have thoughts about home. It's very useful for me to stay here, but it's not that easy. The only thought which makes me feel better is that I'm here temporarily."

I realized that my monastic life would end when I would be able to learn how to be in the state of love - where there are no complaints and no grievances - just as it was on the island. It was easier than before, though at the same time I felt tired from the ascetic life. Sometimes, I was full of contradictions. One part of me accepted everything what was happening with gratitude and for the opportunity of a spiritual development, but the other side wanted it to end. The condition that was made by the old monk confused me. I thought there was nothing else that could confuse me. I had illusions again. It was an insidious provocation of my mind interested in Mohammed's death. "He is old, but he could live another ten years. Am I going to stay here for a decade? No! I want a different life - bright and saturated! I don't want to live as a monk during the best years of my life!", I protested.

For the first time in the last few months I felt a strong emotional discomfort and therefore within myself began to feel a real internal struggle. I thought I was an inch away from the state of integrity, but it turned out that I was still "green", I was still narrow-minded. All these thoughts brought me to the inference, which I stated as an indisputable condition, "Though the monk might die any day, I would pray intensely for his good health and long life." Thanks to his provocations because I had learned the truth.

I realized, "I have lived most of my life being blind. Now I can see the light, but not the objects. There is no point in returning home being half-blind. I will not get out of here until I see everything clear. I will work even harder." I created such an internal statement that the internal struggle waned. Soon, it ceased altogether. My sober mind and strong spirit dominated over the weak side and my mechanical mind.

I overcame the confusions in one day, and I felt as if I had risen up by stepping over my illusions, ignorance, and weakness. I was grateful to the brother Mohammed for this provocation. It felt as if a healer had pulled out of my seemingly perfect body a dark entity which was lurking deep inside of me. Mohammed made this possible by provoking this dark entity into raising its poisonous head. After that internal revolution, I manifested a sense of might over my imperfections. I felt an absolute ease, freedom, and independence from any of my internal negative manifestations. I began to receive a true knowledge which started to visit me during my prayers. My old and boring prayers were transformed into a meditation - a deep state of detachment from the impermanence of this vain world, and I had contact with the true divinity. I did not need to beg God for anything. My goal was to reach a state of serenity. It was not important to me anymore when I would leave the monastery. It became not important for me how long the old monk would live and how long I would live in my physical body.

I understood one more vital wisdom, "*It is important how I live in this particular moment of my life, which has granted to me.*" Gradually, I started to get the answers that I was looking for.

* * *

I pondered, "Is the purpose of my life to improve the Earth? How can I improve it if it's perfect already as a creation of God? It cannot be that we - people, don't have an opportunity to improve what God has created. I don't believe in the theory that God made all for us – his children, and left nothing to improve, so we just have to consume only. I don't believe in it because neither my mind nor feelings agreed with such an idea. So what can people create on such a beautiful paradise like the planet Earth? The only answer was to leave something good after we die! We don't have to improve what's already perfect, but we have to save it and by being images of God, we must become creators – creators of our souls. We are given the opportunity to create a space of love on this wonderful planet with our thoughts and actions. The beginning of this creation is the internal world of every person! Those monks, whose bones radiate good energy after the death of their body explained what state of love they had achieved in their life and what space they had created around themselves. The main point - everyone can find a power of Creation and to reveal a divine love within."

These thoughts came to me when I was going to bed after the evening prayer. I walked slowly looking down at my feet and thinking deeply. Then I felt a touch on my shoulder. I turned around and saw Mohammed who smiled graciously.

"My congratulations brother! Tomorrow you will leave the monastery. You have to go."

"Why? Are you leaving tomorrow brother?"

"Yes," said the old man solemnly. "I have to go as well, but we have different roads. I've completed my journey, but you, at your new level, are just beginning it."

"I will stay until you are buried to show my respect."

"It won't be beneficial for you and definitely not for me. To see the funeral of *my body* isn't a great mission. To remember my soul, you can do anywhere without contemplating my "shell" which tomorrow will become trash just like a trash that you cleaned up after the tourists. In the regular world it's normal for people to mourn about the trash – the unsuitable shell which was left their relative or friend, but actually it's the soul celebrating another birthday. When a soul is freed from a body, it celebrates the return "home". Therefore, who has left doesn't share the grief of those who remained on the Earth. You successfully got rid of the silly tradition of comparing a dead body with those who you love. You shouldn't linger because you have to go right away. You have to do what you are supposed to do. You successfully passed all the tests at this point. You have opened your heart and you have become spiritually stronger, so now you have to move on. Tomorrow morning you will leave, and you will return to your country. I will give you a last farewell. You will go back, and soon you will get used to it again.

Do not take the usual life of the "normal" people that you will be surrounded by seriously. The phenomenon of the material world is just a game. You play, but make sure your game is clean. Follow the rules and take care of the priorities. You don't have to be in a hurry to catch up. You don't need to worry about how to beat the others. You need to learn how to direct your mind to one point. This point is God. God is omnipresent. God is in you. This is necessary in order to calm your mind by giving to the mind the opportunity to get rid of unwanted thoughts. The success in achieving the real goal is not in increasing the speed of the movement to it, but in peace and purity of your mind. Ask yourself the question, 'Who am I?' and do not let your mind answer it. Do not let a single thought be the answer. When in the response will come the silence, enjoy it. This is the true answer.

Do not rush to understand the meaning of my words now. Keep going, and the realization will come. One more thing. You will certainly do good deeds, but remember the important rules: *When you do something good for someone, forget about it immediately and do not burden your memory with quantities of good deeds.* If someone wants to pay a tribute to you, don't deprive them of that joy. Remember, *'The happy one is not who has everything, but who gives.'* These rules are for the Great Spirit which you are to become. I was happy to know you. Goodbye."

We embraced. I looked into the eyes of the old man - they emitted happiness. I felt his love. He blinked his eyes in farewell, turned, and walked away.

I did not feel any particular emotion about my departure. Just in time I was ready for something new. I was ready for a new phase of my life. I felt solemn about Mohammed's life coming to an end. It was so unusual that I had a state of joy for him. In fact, I felt a joy about what is traditionally supposed to be sad in "normal" society. The old monk gave me a valuable lesson which destroyed my stereotypical attitudes towards life and death. I was delighted because I had discovered the truth that was hidden under the veil of the ancient traditions which were generated by ignorance.

CHAPTER 6

Flight Sharm El Sheikh - Moscow. The tourists are returning home - to Russia. I had not heard my native language for so long! I was very happy to enjoy the other passengers' conversations when they were looking for their seats. I love you so much my dear fellow countrymen. You cannot even imagine!

I was observing myself with interest. I was in a familiar secular situation - sitting in the airliner's chair, and in my inner state I felt an extraordinary comfort, peace of mind, and confidence. In my memory was floating the conversation with Dianand. I was so grateful that I had met him. All the gold in the world was not worth what I had gained from that meeting.

"How great it was that he sent me to the monastery where I learned the important lessons and polished what Ilistre, Tare Mugu, and Dianand had given me." I remembered Ilistre without the slightest trace of feeling of loss, but only with gratitude and affection. There was a strong feeling of her presence and the presence of her father Dianand. I recalled how on the island I felt like I was a student because Ilistre knew what was unknown to me. At the same time she always let me understand that I was a real man who she adored. At the monastery I was smoothed with a file by an invisible hand which had been patiently and carefully removing all the burrs and roughness from my nature for three months. I gratefully recalled the monastery which had become my abode, home, and school. The blessed place where for the whole history of its existence monks managed to avoid looting and to preserve priceless relics such as books and manuscripts of the 6th century that were left from the Russian church. The monastery is located on the same territory as a Muslim mosque and it is an illustration of the fact that God is love, and there is no animosity.

I contemplated, "Today, Mohammed will have to hold the answer for the way he lived his life. I don't know his biography, I don't know how long he stayed in the monastery, and I don't know why he was there. Perhaps, it's not so important. What is important is that he will die happy as a student who graduated from an academy far away from home and will soon return to his parents."

Then, I remembered Dianand's words, "There is no tragedy in death, but there is a personal tragedy of the soul when the soul within the time limit of the contract has not fulfilled its purpose."

I continued to ponder, "Apparently, Mohammed is sure that he will please God. He knows that he completed the tasks of his soul, so he is not afraid of death. Death for him is a deliberate withdrawal from the world in which he came as unconscious infant.

What a great feeling that I didn't succumb to my weakness and run away. Otherwise, indeed, it would have been a tragedy. It's awful to be weak because there is a chance to miss so many opportunities in life and never understand what causes trouble."

The plane taxied to the runway, accelerated, and soared into the sky. I was lost in my memories and soon fell asleep.

CHAPTER 7

"Everybody sit still! If here are heroes who aren't afraid to die, stand up, but you won't stand for long. Ha-ha-ha. Today is our day. We are terrorists and we've seized the plane, and you are our hostages." I heard the words through my drowsiness and the hum of turbines.

When I opened my eyes, I saw two men. One had gray hair and the other one had red hair and a very ugly face with close-set eyes. They were walking back and forth while holding guns. How had they carried guns on board? It should have been impossible. Weapon detectors were everywhere. Maybe they were plastic. From the first cabin came an unusual noise. Perhaps something was going on there as well. Where had they come from?

"Dear passengers, did you know that all of you are idiots? Ha-ha-ha." Said gray-headed man.

"Nah, Gray. They aren't idiots, they're flying retardes. Ha-ha-ha." The red-headed man sneered and they both broke into hysterical laughter.

"These terrorists look more like stoned teenagers, although both of them are around thirty." I thought.

Then one old man spoke up, "Young people, don't you think ..."

"What...?!" Exclaimed 'Gray', as if he was surprised.

"Here is the first victim." 'Red' grinned. He was standing next to the chair where the old man was seated. He put a gun to the man's forehead. The old man closed his eyes and cringed in fear. 'Red' moved the gun away slightly and fired. The loud shot caused my ears to ring. Through the cabin light smoke and the smell of gunpowder was spreading.

"That's not a plastic gun", I thought feeling my pulse increase.

They were morons. Messing around with a gun on a plane at an altitude of ten thousand meters, it's equivalent to playing Russian roulette - a readiness to commit suicide. 'Red' fired towards the tail of the plane, not to the side, so at least he has some brains, but he could have hit a person. He didn't care if someone might be there. They were totally indifferent to the lives of everyone on the plane.

"What do they want? What can I do?" I started to panic. "First, I should calm down and bring my pulse back to normal, otherwise nothing would go right. Just a few minutes ago I was in a state of tranquility and pleasure and all of a sudden I'm panicking." It meant that circumstances and people could still control my state. So, I wasn't the absolute master of my emotions and my life yet. I should be able to figure something out. Any problem had a solution.

I remembered Dianand's words, "There is no tragedy in death". I felt the absolute composure to the situation, and at the same time a sense of a presence and support of Ilistre and her father.

"Hey dodderer, you got something to say? Do you hear me, old buffer? Do you have any questions?" 'Red' frowned at the man. His grimace showed a hate and his face became maroon in colour. "What did you say? Are you shivering, scumbag? If you say something again, your brains will be all over the plane!"

The man covered his head with his hands, closed eyes, cringed, and sobbed as he shook.

From the moment the gun fired I felt as if I was electrified and then the necessity for action began to grow. My emotions and heart rate began to calm down and my mind started to think quickly, "These guys have gone too far. I have to stop 'Red', but how? I do not know. I can't watch this any longer, it's almost better to get myself killed. I must find a solution quickly," I said to myself and suddenly I felt as if I was literally increased in size. My mind pulsated, "Death is not a tragedy."

'Red' huffed, "Listen here idiots! We're your bosses and superiors. If we want - you will live, and if we want you to die - you all will die. It depends on our mood. You got that? We can do whatever we want. Come on, you bitch, come here."

He grabbed one lady who was sitting next to him by the hair, put the gun in his belt, and began to unzip his jeans. The girl clung to his arm trying to get out of his grasp as hard as she could; her whole body trembled with fear.

"That's it! That's enough!" I yelled to myself. I made the right decision immediately. I threw myself up on my feet because I had no time to think. All this happened in a fraction of a second. I had a decree from above for action. I was unusually calm and confident. At the same moment I felt myself incredibly huge as a mountain. In my state there was no hate but sort of empathy, and in my mind flashed for a moment, "They are so naïve." I felt a huge potential of inexplicable force and extraordinary power coming from within.

While 'Red' was struggling with the girl, 'Gray' controlled the passengers. When I unexpectedly rose up to my full height in the hall, his grin was gone immediately. I noticed that there was surprise and fear in his eyes. Instantly, this terrible terrorist turned into a surprised young boy. 'Red' did not see me. He was ready to hit the girl…

"Look at me!" I said loudly. It seemed that voice wasn't quite my own.

'Red' turned his head, released the girl, and quickly snatched the gun out from his belt.

There were fifteen steps between us. Feeling like a mountain, I took the first step and went towards him. 'Red' pointed his gun at me and screamed hysterically, "Hold it! I'll shoot!"

"You are going to take only three more shots in your life." I answered confidently slowly approaching him. I could clearly see that his face was contorted with anger. My actions were definitely not a part of his plans. After the first shot (which had been used as a psychological attack), most people would have been in shock but not me. From his point of view all the passengers should be greatly frightened and stay quiet for a long time. I moved towards him looking right into his eyes, and I saw horror in them. It felt as if I was in a slow motion, and I knew everything in advance. With a mild surprise I noticed how the events were unfolding. I saw how 'Red' was aiming at me, and I saw how his hand was shaking in anger. I felt only pity for this wretched man before me who believed that he, when he wanted, could stop the life of any person on the basis that they were a hindrance to him. I saw how he pulled the trigger. "Maybe I'll see you today, my brother Mohammed." Flashed in my mind.

Bang, bang… Again there was a loud din, but this time it merged with the screams of frightened women. Surprisingly, the children who were on board never made a sound. Bullets passed me and with resounding blows punched hard the aluminum bulkhead behind my back. Then there was the third shot. At this point, 'Red's head twitched back sharply, his body jerked, and the bandit awkwardly fell back bending his legs under him. His face turned blue, and the right eye was bleeding badly. His hands moved slowly, his body twisted, and somewhere from his chest came out a groan. In general, it was a picture of the terrible torment and unbearable pain that man can suffer when he was in the semi-conscious state of the traumatic shock.

"These guys are amateurs. Perhaps, they bought the weapons from a black market." Came to my mind. Apparently, the moment the gun fired it damaged itself, and a piece of a chipped metal with a speed of a bullet hit him right in the eye. What was the purpose of all this chaos?

I walked right up to 'Red' who was lying on the floor. Puzzled 'Gray' was looking at me with a pointed gun at my chest.

"Come here." I said quietly staring in his eyes feeling that my eyes were as lightning that could burn him to ashes.

"Do you want to kill me?" I asked calmly. 'Gray' negatively shook his head.

"Don't play with weapons. Give it to me and pick up your buddy's gun too." I stuck the index finger out of my left hand. 'Gray' hung his gun on my finger and reached for his partner's gun. He kept looking in my eyes as he slowly took the gun which had just destroyed all their plans.

"Guys, who did you whack there? I haven't decided yet who I should start with." We heard a cheerful voice from the first class compartment of the plane. 'Gray' looked at me questionably trying to ask what to do now.

"Take your friend and bring him to the entrance door." I quietly commanded. 'Gray' picked 'Red' up under his arms and started to drag him. I followed. We moved into the lobby which was separating two parts of the plane. With a big effort skinny 'Gray' brought his friend to the door, sat down on the floor next to him, and looked at me questioningly again.

At that moment the curtain opened and a third member of their team, this one brunette, appeared in front of me. He was an impressive size of a man with an insolent smile on his sleek face. In an instant the smile disappeared from his lips. He caught a glimpse of the bloody 'Red' who was in unbearable pain, then at 'Gray' sitting on the floor next to him, and then at me with two guns hanging on my finger. He looked perplexed.

"What's going on?" He asked finally.

"That's it buddy. The show is over." I said casually and friendly.

We stood facing each other at arm's length. I could with a sudden move grab the hand that was holding the gun, or I could sharply strike him in a jaw and then disarm him, but I knew none of this was necessary. I felt the power and the extraordinary speed, but not because of an experience of a melee fighting. I had a sense of total control of my body, and I had a complete control over the situation. All of it gave me confidence and to use a fist was a last resort in extreme cases. At that moment the situation was under control, and an extreme case hadn't become. I knew at that moment there was no need to strike. The amazing absolute knowledge of subsequent events held me from unnecessary actions, and I quietly followed the voice of my intuition.

"Put the gun down." I said.

"I'd rather die." Said 'Brown' then raised his arm and pointed the barrel at my stomach.

"Drop the weapons on the floor!" He blurted out panting.

I dropped both guns and they thudded on the floor. I was surprised at my calmness and a sense of own security despite on the obvious danger.

"Put your hands behind your head!" 'Brown' commanded.

"My hands will remain where I choose them to be. Regarding your comment I can say that you have to stay alive for some time. You'll have to go through a lot of pain. In your nightmares the people who you've tormented will come to you and they will scrape your brains with their fingernails asking you why you were so cruel. All the pain that you have caused to others, you will experience yourself. You're going to wish that you were dead to escape the suffering, and you will try to hang yourself, but still it will be too early for you to die."

I looked straight at 'Brown'. The situation was not in my favour, but the feeling of absolute control stayed firmly in my consciousness. I also had knowledge that feeling was not an illusion. Out of the corner of my eye I saw 'Gray' who was frozen in his spot watching.

'Brown' grinned, "So far I have the upper hand. I'll pull the trigger and I'll kill you. I could even just choke you with my hands."

"'Red' already tried that and look what happened to him." I stated.

At that moment, as prove, 'Red' made a heavy groan and arched in an awkward position indicating unbearable agony. 'Brown' was not sure what to do, and I was positive that he would not harm me.

"'Gray', what kind of nonsense is this moron spouting?"

"He's telling the truth. He told 'Red' that his third shot will kill him, so it happened. I think he is a ..." 'Gray' raised his finger and rolled his eyes somewhere upwards. "He isn't afraid of anything, and he knows what's going to happen in advance."

"I'll find it out right now then." 'Brown' smirked.

I felt an incredible speed that I could act with. At that moment inwardly I knew, I had one second. I made a lightning fast motion and the barrel of the third gun rested on his neck under his jaw. With an iron grip I held him with one hand by his wrist and with the other hand by his elbow. I pushed him to the edge of the bulkhead, so he couldn't break away. Because of my sudden move which completely changed his situation, his eyes widened and his eyebrows rose to incredible heights.

"How about now? Do you still have a desire to shoot?" I asked.

The attempt of my sparring partner to break away brought him great pain, so in the effort to get away from the pain 'Brown' sat down. His wrist was turning purple from the grip and his fingers slowly began to unclench. I strongly shook his arm and the gun fell out. A painful grab made 'Brown' to scream.

"Ah-ah! It hurts! Let me go!"

"Hold on a bit. Feel the pain. Next time, this will help you to make a better decision. The rest of the flight you will have to spend on the floor being tied up. Got it?"

"OK, OK! Let me go!"

I let him go and the big guy who had been in charge of the situation just a minute ago fell like a sack of potatoes to the floor beside 'Red'. He was astonished. He glanced at the gun that was lying next to him, but he did not dare to move. 'Brown' was physically very strong and he was sure in his strength and, as always, counted on it but he had suddenly been confronted with more strength, so he was discouraged, dejected, and depressed. He could not comprehend why he had not just shot me.

"Who are you?" He asked.

"Who am I? I'm your punishment. I see that you didn't expect it. We definitely had to meet, so the time has come."

"That's weird. Why didn't I kill you? Why didn't I take a shot?"

"The punishment cannot be avoided. You put your life's force in it. The shooting time is over. Now, the time has come to answer for your deeds. Are you ready to answer?"

"I'll answer." 'Brown' muttered hostilely as he looked at me.

And just then I noticed a flight attendant with bulging eyes standing against the opposite side of the open area. "Agrippina" I read the name on her badge.

"Agrippina, my dear, please bring me some duct tape or something to tie them with." I said to her.

I watched with interest what was happening to me and around me. I could feel the power that was coming from me, or perhaps through me, that made the criminals obey. "Where did it come from?" I wondered.

In that moment in front of me emerged an image of Mohammed. I "heard" a familiar voice of the wise man, *"This is the significant power of the Truth. This is the power of God, the power of the universe which has manifested through a strong spirit. A "bad" person obeys this force as if it was his or her own accord. This force is very powerful, so people are unable to resist it. They are God's creation. Deep inside of their human nature, where the soul is, there is a place which is responsible for the truth. This place in the crucial moments such as this, unconsciously speaks for the truth. A conductor of that power as the strong light shines through the lost soul and then Satan, by being afraid of that light, comes out from a bad person."*

I shouted in my mind, "Mohammed! Brother! It's you! All this time you were here!" I realized that Mohammed was with me all that time, and he had walked me through the corridor of the events. I continued to ponder, "That's it! That's why. People don't die, they just leave their bodies." The old monk just showed me the possibilities of what an angel of light can do. It was hard to understand this all at once because it was so unusual to separate a person's death with the death of the body. Now I know it for sure. I was pierced with the sharp idea, "*People mourn for those who are dead because they think that their loved ones are just bodies.*" For a moment I felt incredibly sad - there were such absurd traditions human societies lived with. Apparently, Mohammed and angels had an impact on the main characters of these events. These criminals couldn't do anything because their abilities couldn't fight the power of the angels who were on my side."

The flight attendant did not have duct tape, so she brought some rope instead. She was still not steady from the experience of horrible events - her hands were trembling and her eyes were intense. I ordered the terrorists, "Sit back to back and put your hands on your elbows".

"Why?" 'Brown' asked. "I want to use the washroom." He added.

"You will go to the washroom with a guard when we land. You guys are unreliable and poorly behaved, so I don't trust you." I answered.

"What do I have to so, wet myself here?" 'Brown' screamed angrily.

"I don't really care. You came into my space with the bad intentions and this came as a result. After all, I didn't ask you to do these things. Perhaps, you have to understand something about yourself. Why did you have to bring yourself to the point of sitting on the floor and soiling yourself? Ask yourself this question. Use your brain. Maybe you will realize that soiling yourself isn't so terrible compared to how you messed up your own life. If you understand that, then you can start thinking about how to clean up what you have done. So, don't whine and do what you have been told. Put your hands on your elbows and I'll tie you up so the passengers will be able to relax. You scared them too much."

"What if I don't agree, then what?" 'Brown' continued, giving me an evil smile. "How about a fight?" He added.

"I don't care whether you agree or not. The result will be the same - you will sit quietly on the floor unable to harass anyone. Don't even think of misbehaving again."

"You're saying that because you have a weapon."

"Yes, you're right. I'm saying that because I have weapons. My weapons are truth and love. All this, gives me the strength against your physical weapons which are useless, even though you keep looking at them hoping for something. You think you're strong when you have a physical weapon, but it's an illusion. Apparently, you cannot figure that out. Look, 'Red' lost his eye thinking the same thing. You're upset because you feel humiliation, but I'm not trying to humiliate you. I have no such purpose. You did everything what was possible to be in this situation. Now, you have to deal with it by yourself. I can only help you by not letting you do even more stupid mistakes than you already have done."

"Well, just try to tie me." Said 'Brown' thinking to fight me.

"I don't want to argue with you. If you're not going to listen to me, you'll leave me no other choice but to break your jaw and ribs. This means that the slightest attempt to move will cause you pain. So, what's it going to be, your choice?"

'Brown' kept looking in my eyes with a great hate, and I was looking at him with eyes full of serenity and compassion. It lasted about a minute. Finally, he resigned himself to the situation, looked down, and reluctantly folded his hands. I quickly tied him and 'Gray' together making sure that the ropes were tied firmly. 'Red' continued to moan. He was still unconscious. I thought, "Soon he will wake and he might start screaming. Perhaps, I should tie him up and tie his mouth as well, so he won't scare the others."

Then, I realized that I hadn't experienced any hatred for them, though they had committed such a serious crime. On the plane, the children could be terrified after the gunfire which could cause them an irreparable trauma. Their mothers were also stressed because of their hopelessness in such a situation. The fathers had experienced humiliation and an incredible helplessness too because they were forced to be passive and not take any action for they could end up dead. It was an objective reality that kept men away from action, and still I did not hate the criminals.

I thought, "Surprisingly, but not long ago I had a goal to get rid of aggression and hatred. It seemed that I had accomplished just that."

I was sitting and thinking about the trio. I felt sorry for them. They were so young. Why did they do all that? I looked up and saw Agrippina and another flight attendant. They were looking at me with admiration.

"Well girls, they won't be causing any more troubles, so you can all relax. Call the pilot and announce to the passengers that everything is alright." I winked at them and added that it would be necessary to bandage 'Red'.

Both girls nodded simultaneously and went back to their duties. I was confident that there were only the three men involved, and there was no one else on board who was part of it.

I removed the cartridges from all three guns. Yes, there were real bullets in them. I saw some marks on one of the barrels from the mechanical grips which damaged the gun. That is why it failed when 'Red' took his third shot. Whoever put this gun in the grip was an idiot. Why had they done it in the first place? It made no sense.

While I was thinking it all, I suddenly felt tired and I had a great desire to return to my seat to take a nap. The powerful energy pouring from me had ceased. "Right on time". It lasted exactly as long as it was needed. Thanks my angels." I thought.

"Hello." I heard a voice from above. I looked up.

"I'm the pilot of the aircraft. My co-pilot is handling the plane. Could you please explain to me what happened here? I was told that the plane was being hijacked."

He was about forty years old. He looked serious, and I tried to put him at ease by saying, "Well, Three boys escaped from kindergarten. Somewhere along the way they found adult bodies, dressed in them, and played terrorists for a bit."

The pilot winced slightly. His pale face indicated that he wasn't ready to joke around, so I apologized and continued, "I really don't understand what they were trying to accomplish. These people were out of control, so I had to interfere, and now you see the result of that. Perhaps, these guys by themselves will tell you what they tried to do here. You can ask the passengers as well. Most importantly, take these guns as evidence. Now would you excuse me? I am terribly tired and I need a nap."

I handed the captain a brand new Berretta as well as the other two guns which had been remade from a gas weapons. He gently took them as he was afraid to catch something from them and said, "Agrippina, please give me a plastic bag."

Agrippina quickly found a bag and the pilot put them away. It was obvious to me that he was confused because he did not know whether to be upset about the situation or happy because it was all over.

I was sitting on a floor and felt that if I did not get up I would fall asleep next to my new "friends".

I heard a noise coming from where I had been sitting. Probably the passengers beginning to recover from their shock and starting to discuss what had just have happened. I struggled to my feet, opened the curtains, and I started walking to my seat. Everybody started clapping their hands (just as the passengers thanked the pilot when the plane successfully landed). I was astonished, "Just a few minutes ago my intuition accurately guided me and I knew everything in advance. A minute ago I planned to take a sweet nap, but my intuition deceived me." I realized that I would not have a chance to sleep.

Then, one nice young lady got up from her seat, quickly came up to me, looked at me with wet eyes full of emotions and without giving me time to react, hugged me. I realized that she was the lady that the 'Red' was trying to rape. No matter how bad I wanted to collapse in my seat, I was very touched. A sweet energy was coming from the lovely lady. She was unusual.

Unexpectedly, I clearly remembered Ilistre and for a moment felt her in my arms. I closed my eyes and tenderly embraced her. "What a pleasure to hug you, my gentle Ilistre", my thoughts wandered.

We stood still for some time. I woke up because the other passengers were excitedly started to denounce the failed terrorists, "Such bustards! They should be in jail for the rest of their lives! They could kill us, scumbags!"

I came back to reality. Ilistre disappeared and in my arms was the unknown cute lady once again. Then, I heard the threats the passengers were making towards the criminals. I was discouraged because the people around me had become judge, jury, and executioner. At that moment, as if by command the children all started to cry at once.

"A few minutes ago I had confronted the aggression and cruelness of the criminals, and now aggression manifests from the good people. Their behaviour is scary. There is so much aggression, revenge, and anger despite the fact that none of the passengers got hurt and the only unpleasant feeling they might have experienced was dirty underwear." I thought.

I looked in the girl's eyes and sincerely hoped that the crowd had not influenced her. I looked into her wet eyes and sighed with a relief - I saw happiness in them. I felt her state. She was as a beautiful lotus among the dirt. Yes, I had a feeling as the negative reaction and the evil energy which was filling the plane hadn't touched her. The "grime" was coming from the good people while I held a beautiful flower in my arms. I had such an association. She had experienced the violence more than anyone on the plane, but she did not hate the bandits. I admired her. At the same time I was confused with the behavior of the other people who were "celebrating" the happy ending. The children, as a fine instrument, showed the imperfections of the adult world and had burst into tears not when the terrorists appeared, but when negative energy from the good people had started to pore out. Apparently, the current situation was scarier for children than the previous one.

After those six months that I had spent in a state of purity on the island and then the three months in prayers in an atmosphere of love at the monastery, I was very sensitive to the negative emotions. It was probably a normal reaction for civilized people, but not for me and not for the children.

"My name is Lena." Said the lady who had a velvet voice. It could be compared to the song of a nightingale. My drowsiness instantly disappeared. I stared in her eyes trying to understand what was going on. I had heard Ilistre's voice! If my eyes were covered, I would have been convinced that it was Ilistre speaking Russian to me. "What's this? What's happening?" I thought, sinking into her eyes. She had beautiful hazel eyes, but her voice was just like Ilistre's.

"Who are you?" I asked perplexedly thinking. "Where did your voice come from?"

"I'm Lena." She repeated, showering me with enjoyment of the familiar intonation of Ilistre's voice.

Still under the impression of the familiar and beloved voice, I thought, "I have to help these people to get rid of the negative emotions. Not too long ago I was like them. I can understand them. They were frightened and felt nervous, and that has led to hatred. I'm in a better frame of mind because I have skills, because I had great teachers who provided me with the knowledge and it gave me strength.."

"Please wait for me a little bit. I'll be back soon, and we will continue." I asked her.

She nodded happily. I went back to the lobby and asked the flight attendant to pass me the microphone and once I had it I began to speak, "Dear friends! My name is Alex. I'm a human being and a passenger just as you are. It seems that we have all become participants of these unpleasant events which are already in the past. These men have lost their minds, and they threatened us with the terrible behavior. Right now, they're not threatening anyone. They allowed me to tie them and now they're in an unenviable position. Their future prospects are also unenviable. All three of them have screwed their lives up by desiring to hurt other people. With them that makes sense because they are criminals, but I'm worried about you. You're not criminals but now you're filled with anger and malice. Please stop! You don't need to be angry. Forgive these people. Don't judge them for their stupidity. Don't hold a grudge against them. Don't poison yourself with their evil. This has to stop. The men are tied up and possibly rethink about their actions, but it turns out that the evil thoughts continue to live. You picked up the banner that was dropped by the terrorists, and now the evil triumphs again - how fast you have been recruited for its service. It's scary. It's terrible that the evil is persisting and how it can manipulates people. In a burst of emotions you forgot that you have compassion, mercy, and kindness. You forget that there are children watching and learning from you. Remember this please: rise above the evil in the name of goodness. Show your compassion and show your mercy for these lost souls even more because no one got hurt. Let's not take on us the heavy burden of judging others. Let's multiply the good. I hope you understand. Please forgive me for preaching. Thank you."

I sighed with a relief and thought, "I expressed my thoughts the best I could. Now, it's all on them."

I turned my head, and I saw 'Gray' looking at me with wide-open eyes full of tears. He was trying to say something, but he could not. He was "heavy" with emotion.

"I'll be happy for you 'Gray' if you can heal your soul in this life time. You have a lot of work to do. Good luck to you." I sincerely wished him. I thought, "He's not hopeless. God bless him."

'Brown' was severely nauseated and his face was green. Apparently, he was suffering from his new condition that made him physically ill, and probably did not hear anything because he was in too much pain.

"What can you do?" I pondered, sighing. "What's done is done." I did not have any pity, sympathy, or joy. I just had some dizziness because I was terribly tired. Also, I had a feeling of satisfaction because I had done everything I could for the people on the plane: the passengers, crew, criminals, and Lena. Then I realized, "No, wait a minute. For Lena - not yet. I still have to talk to her. I still have to figure out why she has Ilistre's voice."

Despite everything, I didn't experience triumph, temptation, or superiority as Dianand had warned me about. I talked to myself, "Thank God. Or maybe I'm fooling myself? Why had I wanted to seal the 'Red's' mouth? Well, the main thing is that I'm aware of these things... If they only knew how badly I wanted to sleep..."

At that moment Agrippina approached me. It was evident that she was beginning to recover from all what has happened – she was noticeably brighter and her cheeks were pink again. "There are a couple of seats available in business class. You won't be disturbed, so you can have a good rest there." She said.

"Thanks Agrippina. You're so nice. You have a rare name."

"Yes." She replied with a smile. "I got it in an honor of my grandmother. She was very wise woman. Come with me and I'll show you where you and your girlfriend can rest."

"What are you talking about, Agrippina? She is not my girlfriend."

"She is. I Know. You have the same color of aura. Please sit here and I'll call her."

"Did you learn that from your grandmother, Clairvoyance?"

"I've learned something."

I lounged in a comfortable chair and luxuriously stretched as I closed my eyes and thought, "With what a pleasure I'll sleep now. How surprisingly the events unfold. I cannot believe this is happening to me."

Lena set down next to me, gently as a cat. I turned to her. The tenderness itself in the most vivid manifestation was looking at me. She took my hand and asked me, "May I sit with you? I'm still a little sore. I feel safe next to you."

"Be my guest. I'll take a nap if you don't mind."

"Not at all, Alex."

I shivered. She said "Alex" just as Ilistre did. The intonation was the same! What was that? I intently looked at her trying to see Ilistre in her, but Lena looked different. I shook my head and smiled. "Now, we're acquainted." I said.

Lena reached out to me and put her cheek on my shoulder. I was very confused as I fell asleep…

I dreamed about of a carousel of constantly changing events. First, there was a sort of horse race, and I was chasing someone in a white-guard uniform. I was aiming and shooting, but from the barrel came jelly that stuck on my blue breeches. Later, I was on an airship flying over the ocean. Under me were many small islands with palm trees, and I could not choose where to land. Then it was winter time and I was skiing in the woods with a gun behind my back. I was a hero in a guerrilla unit. Finally, I dreamed of Ilistre. She was telling me, "I didn't leave you. I'm with you. You're a warrior of the Light." Then I went back to the guerrilla group again. A bearded guerrilla in a white coat with a yellow halo and a purple trim around his head started to shake my hand and drag me somewhere...

I woke up. Apparently, I slept for a short time although I slept well enough. The plane was shaking badly and Lena strongly clutched to my arm.

"Lena, what's happening?"

"I'm afraid. I'm scared."

"Well, before you said that you felt safe with me."

"Aren't you afraid?"

"It depends of what. There is probably something that I'm afraid of, though I can't remember right now."

"Do you feel how the plane is shaking? It seems to me that during the shots something got damaged and now we might crush."

"Relax Lena." I yawned. "Nothing is damaged. The shaking will stop in five minutes. This is just the turbulence. There are rough roads in the air just as on the ground."

As soon as I said that we heard a pleasant voice of Agrippina, "Dear passengers, our plane is in a zone of turbulence. Please take your seats and fasten your seat belts. Bear with us for five minutes and it will stop."

"You're a clairvoyant."

"That's right. I'm a shaman." I laughed it off. "Are you calmed now?"

"A little bit."

"All right."

"Oh, I'm sorry that I woke you up."

"No worries, I slept enough."

The plane stopped shaking, so Lena relaxed and smiled; however, I noticed that she was still worried about something.

"How are you feeling?" I asked.

"You know, when they announced themselves as terrorists, I wasn't scared. It seemed like a bad joke. When 'Red' fired the gun and started to yell at the man in front of me, I was terrified, and I knew that he would molest me. I nearly died from fear when he grabbed me by the hair and began insulting me. Then I heard your strong voice. I felt so much gratitude and happiness when I heard it! Then again there were the gunshots. I was shocked. I was horrified! When I heard your voice again, I cheered up. I was so worried about you when you were behind the curtains. I barely restrained myself from running after you. I remained in my seat because I knew that I could make things worse. You were in there for ages. I thought about only one thing – when you would come back. If you only knew for how long you were gone… I said many prayers for your safety. Then curtains parted and you came back in as if nothing had happened. You looked tired and a little bit funny. Joyful tears came to my eyes and I couldn't hold it in. I rushed to you. When I approached, you gave me a hug and I experienced something amazing and beautiful. I felt your declarations of love to me. Maybe it's silly, but that's how it seemed to me. I felt a powerful energy of love that was coming from you. I wanted to dissolve in it. It seemed to me that we were not on the plane, but on a cloud. I have never experienced a higher pleasure than being embraced by you. I was happy just to be with you. Then there was your speech. It was fabulous! I admire you! You have so much generosity. 'Red' shot at you. You could have been killed, but you don't hate them. Later, I felt myself as a little and silly girl when Agrippina called me to see you. Then again you were with me. Again, I was bathed in the waves of the strength and tenderness that was coming from you. You fell asleep and I watched you sleep with pleasure, and I could not believe that people like you exist. Now you are next to me again and I can touch you. Is it a dream? I hope it's not. At one point I was stung by a thought, 'What if you hadn't gotten on this plane?' It would have been awful. Or, 'what if you had gotten killed?' I felt so bad. I was scared because there is so much evil in the world. When the plane started to shake, I was even more scared. I didn't panic because you were here, beside me. I held your hand tightly, and I woke you up. I'm silly, right?"

"Don't say that, please. No, you're not silly. You're smart. You went through such a trauma today and any person would be stressed. Don't worry about what might have happened. It's all over. Everything is well, thank God. Don't think about the past and enjoy the present. I'll tell you one thing I'm very pleased that I've met you."

"Really?" She said with joy and relief. "It feels so good to be close to you. You just said a few words and I feel peaceful again. What about you? What are you made of? You went through so many things today. I just cannot understand how you handled it? I don't think you are with Special Forces. You don't look like a military man. Your eyes are different. You look more like a romantic or a philosopher."

"You're very clever. You can see to the heart of things. I don't belong to the military. Indeed, I'm rather more like a romantic or a philosopher. How did you figure it out so quickly?"

"Military people don't philosophize like you because they are trained to kill. I think I know everything about you. I have a feeling that you're not a soldier.

I cannot fit one thing in my head: to challenge the villains certainly can many men, but you disarmed them almost effortlessly. You talked to the 'Red' as if he was holding a balloon rather than a gun which was aiming to your chest. How did you do it? How can so many things fit in one person?"

"You won't believe me, but I don't know."

"Are you kidding me?"

"Believe me. I'm curious too about how it all happened. I've never been in a situation like this."

"You behaved as if you had at least one hundred lives. You weren't concerned about whether they killed you or not. Tell me, how is that possible?"

"Yes, you're right. I was taught not to be afraid of death and, apparently, it has helped me. Furthermore, I'm not alone. I have powerful helpers - my teachers. That's why."

"Was the speech to the passengers also your first?"

"That's right. It was. In fact, what I have said is my position on life today. I really think this way. It turns out that I was lucky to meet teachers who taught me to have compassion. To feel compassion towards the bad guys gives me strength. I don't know how it all happened. What's the difference? I believe in God. So, God helped me."

"You believe in God... You know, while you were away, it was already clear to me that everything would be fine. The pilot was here and he asked if everyone was OK. Then the flight attendances carried drinks asking if anybody needed anything. People began to discuss what had happened. Tell me is it really scary when someone is shooting at you?"

"I think it's very scary for a normal person, but I wasn't scared. I was unusually calm, and I knew in advance that no one would hurt me and everything would be fine. I felt as if I was protected with the armor of a Guardian Angel.""

"So, who are you?"

"Who am I? I think I became Batman for five minutes and now I'm the same as I always was."

"You like to joke, that's for sure, but I'm serious."

"I don't want to be serious. Why should you?"

"I think that I'm falling in love with you. So I want to know."

"You better figure out first whether you think that you're in love or you're in love."

"You cannot be serious, can you?"

"I prefer to talk about serious things with a smile. You're very attractive, and I like you. Love is a wonderful feeling! Amazing feeling! Does it really matter who you're in love with? Enjoy the state."

"What do you mean, it doesn't matter?"

"When you're in love because of something, it's usually a myth. This kind of love quickly passes when the myth is gone. It also happens that love disappears earlier if you can't figure out who he is, so there's no need to worry about it. And if you love at the behest of your soul, then just love him and be joyful."

"It's interesting how you see it. You don't want to talk about yourself, and you ignore my questions."

"Not at all. I have nothing to hide from you. You said that you know everything about me."

"Of course I want to know more about you. Maybe I know something about you, but I want to hear it from you. Probably, you're right. What's the difference? I'll enjoy your company while I have this opportunity."

★ ★ ★

"Excuse me Alex, but one of the hijackers is asking to talk to you. What should I tell him?" The other flight attendant, named Marina, said pointing toward the lobby where the criminals were.

"Don't worry, I'll be right there."

"OK, I'll tell him."

"Well, I have to listen to a confession, just like a pastor. Would you let me go?" I asked Lena.

Lena reluctantly released my hand and pulled her knees to her chest.

"Sure, go ahead. I'll wait for you."

I got up and went to the lobby. 'Red' was sitting motionless and his bandaged head leaning to one side. According to the often-throbbing vein on the neck, he was still alive. 'Brown' was dozing. His head was on this chest and his face was sour and greenish. 'Gray' met me with an intent look. I squatted down beside him.

"Speak." I said.

"Alex. Listen to me, please…" He paused to gather his thoughts.

"Go ahead." I encouraged him.

"I was thinking a lot while sitting here. Before, I thought that all good people are stupid because they're weak. I've always liked to laugh at them. I was sure that evil was stronger than the good, until now. I'm shocked. It turns out that I was wrong. You showed me the power of good. I realized that there is the power of good and power of bad. I understood that you're, uh..., were elected by the forces of good. Something has entered in you. I think I saw it. It was some kind of angel or some kind of force. When you got up, I saw that you were glowing. I've never seen anything like that. Maybe that's why I gave up so easily. Then something came out of me. Maybe it was the devil. Suddenly, I started to realize what I had done to my life and the lives of others. Your speech to the passengers really touched me. Nothing could make me cry, but you changed that. I realize that now I'll be locked up for a long time, but I'm not afraid of that. I'm very disgusted with my life. I don't want the devil to come back to me. Therefore, I would like to ask you to help me. I need you to be with me ... No, no ... I mean to stay in touch with me. I need a connection, so the devil will never come back. Please, will you do it?"

"All right."

"Thanks God. I know you won't lie to me."

"Gray, since when did you became a believer?"

"My name is Sergey. I was just thinking about the "believers". I know many guys who have crosses on chains. The cross and a finger-ring is all about showing off among gangsters. I realized that God isn't on the cross, but in me. To do evil is a dumb thing, and God won't accept it, that's for sure. Yes, now I would like to become a believer. Maybe in a prison where I'll be soon, I'll find a library with esoteric books."

"Good idea. Perhaps, you won't go to prison."

"Yeah, right, I'll be locked up alright. But I don't care. To me there's no big difference either I'll be in prison or not."

"I do not think being behind the bars is normal. Perhaps, you have forgotten how to intelligently handle freedom?"

"Yes, maybe so, but sometimes freedom is worse than prison."

"It's better to learn how to properly manage your own freedom and life. Do you think you can learn it if you really want to?"

"I think so and I believe you speak the truth."

"Sergey, I still don't understand. What did you want to get from the hijacking?"

"We just got out ... you know... from lockup... and 'Brown'..."

Sergey nodded at 'Brown', sighed, and continued, "He offered us a job. We supposed to get big cash and we'd get to chill out at a pimped out resort at the same time. This big dealer was doing it because his wife had gone with his buddy to relax in Egypt. He was losing his mind over it, so he wanted to figure out how to get her back and how to punish his buddy. So, we came to a place where they partied. The idea was to humiliate his buddy while the woman watched. This dealer hoped that after that, his wife would forget his buddy and he would meet her at the airport to get her back. The husband had a perfect plan but it didn't work out coz we screwed up and ran out of time. You know, we went wild at the resort after being locked up so long in Magadan. This big shot dealer has connections, so he made another plan with someone from the airport. He told Brown make a play in business class, where his buddy and his wife were while Red and I had to control the others. In short, just like a Hollywood movie... We were supposed to land somewhere in Pakistan or Afghanistan, where we would be picked up. That dealer bought us for dope. He provided everything. We were jonesing bad before the flight. All the crack was up in smoke, and we couldn't get more till we landed. Such bullshit..."

From this muddled explanation I understood only one thing: the corruption was flourishing because there was a deficit of knowledge about true values, and because of that there was too much stupidity in the world. It turned out that the suffering existed from spiritual poverty.

"Well, Sergey, talk to you later." I sighed.

"Bye. I'll see you at the hearing."

Sergey's eyes looked friendly. He was pleased with the conversation. 'Brown' was still gloomy.

From detectives stories I knew that criminals were fine actors and can play any role so anybody would believe it. What's the difference? *The learning process depends on the person and the result - on God.* Earlier I had resolutely stood up from the seat to accept any challenge of my fate. I was able to break free from the traditional fear of death. Still, I love life. Since life goes on, I needed to live according to my conscience. Therefore, I did not think to deny Sergey's request to support him. Then I went back to Lena.

"Are you okay?"

"Yes."

"What did he want?"

"He wanted to share his thoughts with me. How are you?"

"All right. I'm slowly digesting what happened. As you said I'm trying to think positively. It works. Tell me about yourself. Where is your home? Obviously, you're not from Moscow."

"You're right, I'm not. How about you?"

"Yes, I live in Moscow. I was visiting my friends in Cairo and after that went on a vacation to the Red Sea. Now, I am going back home. I'm afraid that after we land I won't see you again. I beg you to stay in Moscow for a little while. Please don't disappear. I have lots to tell you. It's very important."

"Lena, don't worry. I won't leave you. I have a feeling that I'll be asked to stay in Moscow for some time."

"That's good. Do you have a family?"

"I'm free."

"Why? Every woman dreams of a man like you."

"A man like me, really? Just imagine that none of this had happened, and I tried to get acquainted with you on the plane or at the airport. You wouldn't have talked to me. After all, you're a very attractive girl, and certainly many men are interested in you. What do you see in these men? Can you randomly determine who the potential Batman is? Can you see what are they made of?"

"I guess you're right. I cannot see, but why are you alone? Women dream about men just like you. They want to raise children and have family with men like you."

"One more time: I'm not alone. I'm free and that's what I like. Not every woman would like that. You saw me as a hero. You created a bright and not very objective portrait. This happens pretty often. Time goes by, and to this not objectivity portrait later adds some details which aren't visible at first - and this portrait becomes not so attractive anymore. Later a disappointment comes – it's not what you have expected and you feel as if you have been deceived. Finally, the admiration would be replaced with reproach and complaints. If you're admiring me today it doesn't mean that you will admire me always. It was a shiny candy wrapper that attracted you, not me. It happened that I was in the shiny wrapper. In mundane life situations you might dislike me and my heroic past won't compensate my daily oddities."

"You talk like a sage. It seems to me that with all your qualities the major drawbacks can be forgiven."

"So, would you like to do an experiment?"

"Sure." Joyfully replied Lena.

"Do you have anybody?"

"Yes and no."

"Hm-hmm ... I think I understand what you mean."

"There is a man who fills the emptiness, but I don't have a man who I would like to have children with, and who I want to be around all my life. How about you? Do you have anybody?"

"Maybe."

"What do you mean by that?"

"It happened that I suddenly left my home for almost a year. I left home as one person and now I'm coming back as a completely different person. I don't know what is going on at home, I don't know what will be the reaction of those who I had left so abruptly, and I don't worry about it. I'm confident that I can build my life the way I want."

"Is that so? Are you inclined to disappear suddenly for a year?"

"I told you. I'm a troublemaker, but you thought I was a superman."

"You're joking again. Don't you want someone who adores you? Don't you want to have a family? What about marriage?"

"Before, I wanted many different things, but everything went pretty haphazard because I had no idea what love was. Much later, when my fate faced me with some events, I changed my perceptions. Now, the conventional understanding of love and marriage is unacceptable for me. Often, the relationships between a man and a woman develops following the pattern: they meet, get infatuated and want to live together for the rest of their lives. So they get married thinking to create a strong family but later don't notice how their marriage is falling apart. It is because they weren't ready. As time passed, love disappeared and they began to suffer. In the beginning they were attached to each other with tender and vibrant feelings, but later they were attached to each other because they had to manage their household. This is not the union of loving hearts, but cooperative living. Some time passes, and the inept builders of their happiness don't love each other, but endure. They need each other to make their lives work - she works, goes shopping, cooks for him and the children, and cleans their house; he works, fixes things (if he is capable of doing it), and moves furniture. Everything seems to be in place."

Once I learned what love was, how colourful it could be, how much poetry and delight was in it, how I could enjoy giving to a woman and caring about her as much as I could, things like cooking, laundry, and other general household chores have lost their importance to me. The old and traditional life has lost its value for me.

Relationships where there is no God, no spiritual and internal growth, and no romance but where there is conditionality, framework, and home responsibilities - do not attract me what-so-ever. Therefore, I don't know if my family and friends will accept my current views of relationships between men and women.

Moreover, a desire to be "idolized" is a trap. A woman will idolize me not when I desperately need it, but when I'm capable of making her happy and receiving pleasure from doing it.

Before, I was more anxious to find the one who would be the most useful to me in the household activities. My priorities had changed. Now, I have a healthy need to serve a woman. Now, my necessity to give is considerably greater than the need to receive. I receive pleasure from taking care of my beloved, and whether or not a woman would idolized me, is only a concern of my ego and therefore is unimportant. A person, who is captured by ego, would suffer from typical narrow-minded conflicts and losses. I will not allow my enjoyment of life to depend on how a woman treats me. The main thing is that a woman kindles the desire in me to take care of her. If I decided to have a relationship with a woman with this intent, I would have to make sure that she needs my care. If she does, the relationship would continue to develop. If not, it would fail. At the same time, to tell the truth, I'm still not fully adapted to this world from which I was away for a long time."

"We have a contract for an experiment. Do you remember?"

"Is the experiment for a week? Well, I'll give you the opportunity to leave this idea. You idolize me, but idolization usually fades quickly."

"We'll see. You're a guest in Moscow stuck in the net of a local girl who is going to be your guide. All right?"

"This has happened to me before. My guide and adviser on the island was my lovely Ilistre. Now, it happens again - déjà vu. What does this mean?" I pondered.

"All right." I raised my hands showing her that I was giving in.

"From the airport we're going straight to my place. Later, I'll give you a tour of the capital, and objections won't be accepted." She happily replied.

"I don't have any other choice but obey to the local girl from the capital of Russia, but I think it will be hard to leave the airport because the police won't let me go. They might take me somewhere to ask questions."

"Doesn't matter. I won't let you go either, I'll go with you."

"Great."

"You know, while you were sleeping, one VIP passenger approached me and asked if he could disturb you. He was a little bit drunk, so he wanted to drink an expensive cognac with you. He thought that I was your wife and I would know your taste. I liked that. I really enjoyed the role, and I quickly got used to it. I said that there was no need to wake you because you don't drink at all. Am I right?"

"Whaaat?! What have you done?! How could you do this to me? This is an outrage! You deprived me of the pleasure of drinking an expensive cognac for free! I'll never forgive you. This is the first and last time we fight. I've warned you. The experiment is over. I cannot imagine how you can redeem yourself in my eyes after such misconduct!"

"Phew. You scared me! I'm still not used to your humor. You're so sweet. I'll think about how to fix that. Ohh yes, now I know! That's so romantic, and surely you'll enjoy it. My neighbor, who lives above my apartment, makes moonshine. Maybe I'll buy twenty liters of her nectar and when you're in the bathtub, I'll pour it on you. It will be even more enjoyable for you because you'll be washed with it and you can drink it as much as you want. And it is also for free. Am I redeemed?"

"You're a very clever and resourceful girl. Actually, I would never even dream of such a pleasure. Yes, that's Moscow! Please don't deceive me."

"Dear passengers. We're going to land soon. Please fasten your seat belts and bring your seats to their vertical position. The temperature in Moscow is plus eleven degrees." We heard from the speakers.

"I can't wait." Lena sighed.

"Don't rush. Enjoy the moment. Don't you feel good here with me?"

"Yes." Said Lena. She tightly held my arm. Then she added, "Yes of course, but for some reason I want as soon as possible to leave this plane."

★ ★ ★

After a gentle landing, the plane was brought to the parking area. The pilot announced to the passengers to remain in the seats due to the incident that had happened, so police investigators could inspect the plane. The passengers became a little worried, but after a while they calmed down.

The plane door opened and Special Forces came in wearing masks. They took the three men out and started searching the plane in case there were other accomplices.

Later, the detectives came in and they began their work. They had to reconstruct a picture of the events that had happened by drawing the circuit. After some time the Special Forces left the plane and the passengers were allowed to leave as well. Right by the plane there were many buses which were equipped as offices for the detectives. The passengers were giving their statements to the police.

The whole procedure took less time than I had expected. The investigative activities were well organized. After two hours, I was allowed to leave under the one condition – I had to remain in Moscow for at least one week which made Lena very happy. In the connection with this moment, I remembered the heel of my shoe that was torn. "Perhaps, this situation will lead to something important." I concluded.

While we were in the taxi, Lena fell asleep resting her head on my shoulder. I was wondering about the events that had occurred in the last few months of my life.

"Life is an interesting thing. First, I lived as everyone else by gaining life experience. My parents and teachers at school taught me how to be independent and how to solve problems. Then I began to realize that the knowledge I was supplied with, was unsuitable. I made many mistakes using that knowledge and because of it, I was often got uncomfortable in my life." I was thinking about the imperfections of the laws and the lack of the important knowledge and useful traditions in the society.

I recalled when I was physically strong, and I didn't have enough joy in my life. Then, I began to improve my personality. I had a desire to find a formula for the happiness. There was an intention. I became a seeker.

That fateful meeting with Victor at the airport opened my eyes. It would probably be too serious to call him a wise man, but for me, back in those days, he was. He graciously shared his experience with me and taught me how to manage life. It was the equivalent to a sip of water in a deadly desert. I had changed my interests, so I had attracted many positive events in my life. I began to read different esoteric books. I began to think. As a result, my consciousness and life in general had changed for better. At the same time some unnecessary people had left my life, but others came – who were welcomed, positive, and enlightened. People, who did not leave my life started to change for better as well.

Then I remembered the island, Ilistre, and our love. Those stunning spiritual and energy practices had opened me to new sensations and increased the growth of my consciousness. Ilistre's leaving made me rethink my definition of death.

Then I met her father Dianand and went to the monastery where I lived an unusual lifestyle with hard work that allowed me to make new discoveries.

Then finally the situation on the plane. What did I do? How did it happen? Now, it seemed as if it had been a dream or some kind of play with a hidden meaning. Lena's voice was remarkably similar to Ilistre's. What did it mean? Was there such coincidences?

Divine games are truly amazing. There, on the plane, I felt the presence of Ilistre, and it gave me strength. The experience of love, meditations, loss, and the monastery life - all gave me an amazing strength and peace during the events on the plane. I felt fearlessness because I was unbound to my life. Ilistre and all the inhabitants of the mysterious island showed me that there is no tragedy in death. After the physical death of Ilistre, I began to feel her mentally.

It looks as if all the events that had happened before then, were a preparation for the main event on the plane – the "final exam". It seemed as if all the participants in these events - 'Red', 'Gray', 'Brown', and Lena – were actors in a play. They played their roles in order to help me to achieve strength and love.

What if when we get out of the taxi, somebody said, 'Now look there. You're on a hidden camera. It was a practical joke.' And everybody would start laughing – the hijackers, Lena, the pilot, Agrippina, and the Special Forces?!

I wished that it was a joke, but no! The events were real. One thing was for sure – they were not random. Ilistre was my real teacher. She was my teacher of spirituality, wisdom, and love. Before, when I created the intention of finding such a teacher, I imagined that it would be an old Chinese man, a white-haired Tibetan, or an Indian Brahmin, and they would teach me by telling me the parables in a mountainous monastery or in an ashram.

However, it all appeared in a completely different way that I couldn't even dream about. I never thought that my teacher would be a young woman. I would never think that the spiritual practices with her could be combined with intimacy. No, it was not just sex. It was the transformation into a divine conjunction. It was a revelation of divine love through intimacy. Intimacy is one of the facets of love. The true love is a stable internal state which doesn't undergo to any changes. This love cannot be combined with delusions or depend on them.

What Ilistre taught me, gave me the opportunity to disclose a divine potential, and I felt the process within. It was a process of disclosing my inner strength and my divine nature. This process involved successes and failures, friends and foes, good events and adverse circumstances. Ilistre was my mentor on the island, but now she is always with me - in my heart. Lena is a proof of that. There is some kind of a connection between the two women, and that connection is not just the similarity of their voices. It isn't just a coincidence that I've met a woman with Ilistre's voice. Could it be that beautiful Ilistre helped me get rid of the mental trash, and now it was my turn to help Lena? But why am I simplifying it? Why do I allow my mind to subdue the events to logic? Maybe all that had an entirely different meaning and intent from the heavens. Anyway, what is the difference? I have to act according to my principles without betraying myself and everything would be fine. OK, what is next? Do I have no flaws? Did I bring my best qualities up to the required level? Can I grow more?"

My thoughts made me laugh, "Of course, there are no limits to perfection, and there is still much more work."

CHAPTER 8

That evening Lena and I gave in to the freedom of our feelings. Being in a state of universal love, I discovered the amazing frankness and amorousness of her gentle nature. We could not get enough of each other by pouring out our feelings. My untapped reserve of care and tenderness broke out, and Lena accepted it with overflowing feelings towards me. It was wonderful! The best of it was that she felt and understood me totally.

As soon as her hands touched me, we did not talk. We were communicating at the level feelings. I was grateful to Lena for the fact that she accepted the condition when the words are not necessary and where only feelings were needed – from heart to heart and soul to soul. I caught myself thinking that during our intimacy, Lena was for me the embodiment of Ilistre. I was not upset or glad about it. I just enjoyed what was happening to me and how things were going.

The next day Lena passionately began to study me as if I was an encyclopedia. She threw questions at me and she wanted to know absolutely everything down to the smallest details. I told her about how I used to live, then how I sought for answers, and finally how I began to receive what I was looking for. I told her about what had happened to me in the last few months. She was asking me questions again and again. Our conversation transformed into lessons of geography, history, and philosophy. I lost touch with talking to people because of the island and the monastery where we spoke very rarely, so I was shocked how much energy it took. I was able to relax from talking only at night when we were in bed.

* * *

Lena was delighted by the intricacies of the plots that had happened to me. Sometimes, she asked me in details about some event and excitedly said how unique, incomparable, and impossible they were. She was constantly asking me to say what I was thinking about one event or the other. One day, when I finished one story Lena was staring off into the distance and thinking for a long moment while digesting what I had said before she whispered, "I think all that has happened to you is a rare gift from Heaven. What do you think?"

"You're right."

"So any person can receive such a gift?"

"Yes, I think everyone can, if a gift counts as an event. Sometimes a desirable event can destroy a person and something a serious trouble can become a little key to open a door to new opportunities, becoming a springboard in achieving desired goals. Dianand told me one of the main conditions of all people, *"It is important in what mind frame you meet an event."* So there is a need to be ready for any circumstance, but the most important thing is *to grow love from within.*

"Love! Grow love! It seems so simple before going somewhere. The provocations are everywhere. There would always be someone who would try to ruin my day, but I want to learn how you grow love."

"Yes, I know these thoughts and these feelings. I used to struggle with them every day. I used to meet someone daily who would make me angry or frustrated. Then I realized that the problem was not with other people, but with me. I had to change my perceptions. It was necessary to allow everyone to be the way they were and begin to work on myself. After all, *you don't need to have wisdom, a beautiful soul, nor a strong spirit to condemn, criticize, or blame others. There is no particular merit in blaming and criticism. There is no effort in the negative reactions, but there is a passive flow in the free will of circumstances.* I had to change my attitude towards the life around me. *The first step was to understand the necessity of abolishing my anger and condemnation.* Even though it wasn't easy, but when the first step was made and a person firmly stands up for their new perception of the worldview, inevitably the way how to achieve the established goals comes along. Once, I had an idea. It was as a revelation! We must create a mood. In the morning, before leaving the house, we should tune ourselves to love everybody. We have to enjoy every breath we take, every blink of our eyes, and every step we take. We must accept and love everything that exists in this world. We need to accept ourselves as we are with all of our strengths and weaknesses. As a result of this acceptance, we will emerge with the resources for self-creation. When you treat yourself with kindness, then you begin to treat the whole world with kindness. We need to create the willingness to love everything that we see - clouds, trees, every blade of grass, even loud cars, or even a grumpy person or a mischievous teenager.

"Since I used this technique to see the world, nothing could make me angry, upset, or frustrated. I noticed that by being in this state, the world around me became more attractive. I started to improve the technique, and I focused on having fun and taking pleasure from anything that met my eyes. Everything what we saw and felt was one beautiful masterpiece in God's creation. Also, it was important to smile and to keep your back straight. It used to be new and strange to me. A powerful spring repelled me to the former state, but I resisted the force of an old and bad habit by imposing my will on my own imperfections. I worked hard on this technique transforming myself into a better person. I think because I had a desire to transform myself internally, I was able to receive a permit to the schools with intensive trainings: the island, Dianand, and the monastery. Later I had to do the exam on what I had learned during the situation on the plane. Thus, because of my ever-growing spirituality, my chances during the exam rose from zero to one hundred percent."

"Yes, it's hard to accept all of this. This seems to be extraordinary and almost beyond understanding." Lena said.

"It's only a matter of time. If you become observant and if you control your thoughts, emotions, and actions then very soon you will notice that the boundaries of the beyond are expanding. Everything that today seemed incredible and beyond, tomorrow will look natural and attainable.

There is another important thing to remember - a sense of humour is a powerful factor in the creative work on your inner state. There is no need to take this world too seriously."

CHAPTER 9

At the first opportunity I got in touch with Lenar and told him that I had just arrived in Moscow. He was delighted and said that a few days ago he bumped into Arthur who was in Moscow for business. We agreed to meet at our favourite Japanese restaurant.

At the appointed time I entered a cozy restaurant that was decoreted by aquariums along the walls. My friends were already there. When they saw me, they got up to meet me and yelled together, "Oh-oh-oh-oh, the prodigal son has returned!"

We hugged each other.

"Well, we're together again." Lenar said. "Alex, I can barely recognize you. Is that really you?"

"Is it really me? I think I know what you mean. I'll tell you, no, it's not me. You see a different person. But I know for sure that now I'm who I was meant to be." I replied smiling.

"Well, you twisted it nicely. Now I recognize you. Arthur look, Alex looks younger, doesn't he? His eyes are bright and fresh. Alex, you became as mysterious as those natives on the island." Lenar smiled showing all his teeth.

"Okay, we all have changed a little bit" said Arthur, looking at me closely. "But you, my friend, have changed a lot. If a person was away for a year, it means something. Well, tell us everything."

"What do I have to tell? I'm back. Here I'm."

"Come on, we heard what you did on the plane!"

"Nothing special. It just turned out that way. Better if you tell me how you're doing."

"He doesn't want to talk. Well, today we'll make you talk. Lenar, open that bottle!"

"I hate to disappoint you my friends, but I quit drinking. You cannot imagine how glad I'm to see you, and I'm happy without liquor."

"Another abstainer." Said Arthur. "Lenar is taking it easy and now you... Are you coded?"

"I reprogrammed myself."

"Ha-ha. It seems that the new program is good for you. You look so fresh and healthy. I hope just one sip of real "Cinandali" which I personally brought from Georgia will not interfere with your new program. What do you say?"

"OK, I'll just taste it."

"Well, what to tell you about us? I have no special news. Life goes on. Business is growing. Lenar, after visiting the island, acquired unusual literary abilities. He became a humorist-philosopher or rather a parodist. He encroached on our literary classics and started to seek out the discrepancies in their stories. The main idea in his stories is to cultivate generosity, self-control, and intuition - and then you will certainly get what you want."

"Excellent! Lenar, you always have been a humorist, but now you're also a writer." I laughed.

Lenar continued in a serious tone, "A lot of things have happened... After visiting the island, my perceptions of the world have changed. I began to observe, think, reflect, and analyze. For example, one story made me think a lot. I'm going to tell you an interesting story about my neighbors – two senior ladies. They lived in the same apartment building, but with adjacent entrances. They were over seventy years old. They weren't friends, but communicated with each other on a daily basis. Nothing would be interesting in this story except a few details. Those two elderly women lived in the same building for twenty years, but they faced opposite circumstances. One was constantly suffering the presence of drug addicted men who smoked under her door, spit, set her door on fire, or sang all night. Ironically, the other lady had never seen these men. Sometimes, the ladies talked with each other sitting on a bench. One was always wondering about the fantasies of the other, 'Where did she find those addicts? She is probably watching too much television.' The other was no less surprised, 'Why is she pretending that nothing is happening? So many terrible things are happening around us?' Is it possible that one lady was confronted with positive situations and the other with negative, living in the same building for twenty years?! This factor needed to be researched.

"I studied the lady who was suffering from the presence of drug addicts. I defined her psycho-spiritual portrait and interests - she hated young people for the fact that they didn't dress as they should, sang stupid songs, and because they weren't the same as her generation. She also hated rich people because she thought they were all thieves and crooks. She hated doctors because they didn't know anything about her sores and diseases. She hated the government because they didn't care about her problems. The list of people she hated and the reasons was very long. She was interested in gossiping about everything – from the neighbors to the president. She watched TV shows about tragedies and crimes. She always knew where, when, and how many people were killed or raped in a day.

Ok, now. The other lady liked shows for kids, traveling, National Geographic, figure skating, etc. When she watched dances, she couldn't sit still and tried to imitate them as much as she could and she was delighted by beauty. She admired the advanced and talented youth. Not so long ago, she literally lost all her savings because of the money exchange, but she was strong enough to forgive the government. She felt for those, who do evil things because sooner or later they would answer for their deeds. Sometimes she felt sad because she didn't always have access to philosophical literature, but being a good-natured woman she shied away from gossip and rumor of other elderly people where the main themes were judgmental of their actions, social vices, and their diseases. It turns out that good and bad thoughts worked as powerful magnets. These two different ladies lived in one building, but saw things in opposite spectrums. Is it a mystical? Experts say no, it is normal. So, here is a conclusion, *'The thoughts and interests a person has, she or he brings into their life.'*

I began to keep track of my thoughts. I stopped watching movies with violence and reading newspapers that were negative. I tried to think positively. You know, now I sleep better, I feel more relaxed, and so my business has picked up. I don't need to relieve stress with cognac anymore. I started to read philosophical books. They are very interesting. The best part - I'm less upset with this imperfect world because now I have a different outlook at life."

* * *

We had a good time at the restaurant. I was glad to note that I really did not want to drink because I learned how to enjoy my life without alcohol. Lenar did not drink much and took only a few sips. Arthur was drinking at the beginning, but later he started copying Lenar.

I told them all about my adventures, observations, and transformations. They listened to my story with great interest and when I finished, there was a long pause.

"Alex! Wow! That was quite a story! It seems as we only left the island yesterday, but almost a year already has past. In that period you had a chance to experience so much. Now I understand why you talk like a philosopher. Lenar, do you have any idea who Alex is now? He is a real Guru. You know Alex, I'm proud that you're my friend. Look Lenar, why did we leave the island so soon?"

"You said that it was boring there without vodka, and you said that the girls didn't pay any attention to a handsome guy like you."

"Oh, yes. We were so stupid."

"No, you were stupid. It was then when I started to grow." Answered Lenar smiled. "I, for example, during that festival on the island was in an enlightened state for the whole week, and later I got inspired and started to write... Don't worry Arthur, Alex and I will help you to grow. Am I right, Alex?"

"Yes. So Arthur, are you with us?"

"I agree with everything you said. Look how lucky I'm now: one friend is a spiritual mentor and another one is a genius writer."

After we had a good laugh, Lenar leaned over to me and asked, "Listen Alex, can you do me a favour? I have a partner in business. His name is Slava. We've worked together for four years. He is honest, responsible, and punctual, but at the same time categorical, pretentious, and he is becoming more and more negative. As I understand his health has worsened because of it. Lately, he is ill quite often. He likes to cure himself the old fashioned way - through cognac. Doctors told him not to drink alcohol, but he cannot imagine how he can survive without it. He doesn't want to listen to me. I'm not an authority to him. You're another matter. The spiritual teachers had taught you, and now you have skills to help. Would you talk to him, my friend?"

"Ok, no problem."

"That's great. I know he won't refuse to meet you. I told him about you. Is tomorrow a good day for you?"

"Sure."

CHAPTER 10

Slava was slightly arrogant. It was clear to me that he was not a weak person, but he could not afford the luxury for himself to relax and be open. I thought, "He is like a teenager who manifests an affectation." In fact, that affectation - is a mask. He had created an image to hide behind and protect himself from a "frightening" world.

I looked at my companion and in a few seconds I began to feel his inner world. Lenar was right. Slava had a strong personality, but he was destroying himself. The reason was stereotypical thinking. He had inappropriate self-perception and perception of the world.

I concluded that because his life strength was being misused, it was causing him trouble. His troubles were making him angry and creating dissatisfaction. Dissatisfaction was creating an internal conflict which was taking away his life strength, and it was making him exhausted. He was weakened by the constant loss of his life energy, and he was unable to accept the challenge of the moment. This man, being under such pressure was unable to properly respond to the stimuli that life was throwing at him to correct the mistakes. He didn't want to change his perceptions of the world, so the mistakes were becoming chronic. The internal and external conflicts were getting worse. That was why the categorization and pretentiousness had appeared. Hence, the intolerance he had for the mistakes of others and his painful attitude towards any event made him suffer. He suffered because he had devised his own model of the world.

The cause of his bad health was clear to me. I knew how to fix his situation because I had experienced something similar. It was all too clear to me, but how was I supposed to explain it to him?! How was I supposed to convey the truths that were so obvious to me? Anyway, the process had begun, though, and I would do anything in my power to help him.

I met with Slava and we agreed to be straight up with each other.

"Lenar told me about you." Slava began. "Tell me, is it all true?"

"My friend is an honest man, but he is a big dreamer. What do you want to know?"

"You spent almost a year on a spiritual quest learning different techniques from various great sages in different parts of the world. Is that true?"

I thought, "Well Lenar, you couldn't resist embellishing it. But did he embellish it? Wasn't Ilistre a great sage? Weren't Tare Mugu, Dianand, and Mohammed great sages? I didn't realize how lucky and honored I had been to meet those great people. Thanks to Lenar, my eyes were opened to the fact that I hadn't fully appreciated it."

"Yes Slava, it's true."

"Great. Lucky you, I wouldn't mind that."

"Wouldn't mind what?"

"To learn."

"It's a good idea."

"Can you teach me something? There are so many idiots around and my health isn't good anymore."

"Idiots ... health ... Hmmm. How do you think I can teach you?"

"Well, can you give me advice on how I can deal with idiots in my life and maybe you have some special knowledge from the monks and sages on how to restore health?"

"If so, will you follow my advice?"

"Well, if I like it, then yes."

"And if you don't like it, then no?"

"Eh, our conversation isn't going the right way."

I noticed that his "mask" was coming off and starting to show the natural and beautiful essence of who he really was underneath.

"Wait a minute. Don't get angry. It's my personality. Basically, I need help from someone like you. I trust you. I need someone who is wise enough to look at my problems."

"All right. Tell me."

"Lenar says that my health is deteriorating because I have incorrect perceptions of the world. If I will change my perceptions of the world, then my health will improve. I think this is complete nonsense. If I didn't respect him, I would have told him to get out. What do you think?"

"I agree with Lenar's thoughts."

"Hmm. If you think so, I admit that I've missed something. What should I do?"

"You have beliefs and a stance in life. So if something is wrong in your life, it means that your stance or your beliefs are incorrect. Therefore you should change them."

"How?"

"Totally."

"What about my habits and principles?"

"Do you want to be happy and healthy?"

"Yes, I do."

"Then you need to work on new principles and habits before it's too late. But you're not ready yet. When it becomes absolutely unbearable, then I'll be happy to share with you the knowledge that I received from the great teachers. Ok?"

Slava's eyes widened and I saw surprise and protest in them. At that point, his "mask" fell off completely and I saw his candor and willingness for a change.

"I'm ready! I'm ready now! I don't want to wait. Tell me what to do and I'll cooperate."

"Ok, Ok. You've developed habits and principles while you grew up. It's easier to follow them than to give them up. In return there will be training of a new state and new perceptions of the world. It's a difficult and unfamiliar view. What to choose? Masters say, '*when you have a choice and you don't know what to choose, then chose a new way. Possibly, the new will bring some inconveniences or even losses in the beginning. You might regret because you didn't choose the old and familiar way, but it's a trap. It's a provocation. It's a test. You should willingly go through this period and reach a new level of consciousness and a new state.'*

At that level there is a different quality of a physical health and broader opportunities, in other words, a better quality of life. The requirements of course are also higher, but it's worth it for a creative and forever young soul. This is for those, who want to achieve their goals and who want to learn their true essence. This is for a soul who seeks to gain a state of harmony. So, what do you say?"

"Yes, I'm ready."

"What is the thing you worry most about?"

"My health."

"Bad health is a consequence. We won't talk about consequences. We need to get to the root of the problem, right?"

"I agree."

"What else is bothering you?"

"Why are there so many morons and idiots? It drives me crazy when I come across stupidity and cruelty."

"This is an inappropriate worldview and that's why you're not feeling well. It's destroying your health. This worldview fills your mind with flavors which attracts people that you don't like to meet. If you won't change your worldviews, you're going to sink into a swamp of problems."

"So, what to do?"

"Do you want the recipe?"

"Well, yes."

"Too bad for you. I don't give recipes. I'm not a great teacher as those who I've met. I'm still learning. If I give you the recipe, it means to take a responsibility over your life. That's wrong. You should find the way by yourself. I'll help you. I'll give you my opinion on your problem and after that you're on your own. So, what's bothering you?"

"I told you already… Oh, wait a minute, I got it. The problem is in me."

"Bravo! Well done. Now you know the cause of the discomfort in your life – it's not "outside" but "inside". Keep track of it and get rid of your complaints about the outside world. You have to create a new model of perception for the world. What else is bothering you?"

"Hmm. I cannot deal with problems because I take them to heart and that's why I suffer. I get angry at people who give me trouble ... No, no, wrong. It's better to say that I'm vexed because even a small problem frustrates me dramatically. I'm trying to take it easy, but for some reason I can't." He said.

"If you realize it, then you're close to finding the way out of it. Even spiritually strong people sometimes act as "victims" and allow people, circumstances, or events to affect their mood or state. Some people experiencing an unexpected problem would calm themselves down by talking to themselves like, 'this is really nothing compared to other problems that people might have, so why do I worry about it? I don't need it. I must stop worrying about minor problems.'

A strong person who can realize the uselessness of worrying about trifles is capable of creating a new program to ignore the worry. However, someone might be incapable creating the new habit of calmly reacting to unpleasant events. Then, it's time to comprehend the main aspect of life - LOVE. It's necessary to understand the questions: '*What is love*?' and '*What is it to love*?' If you comprehend the truth which shows that love is your inner state, you can start to create this state within yourself. *All the famous Masters, in general, strive to teach people to love others.*'

"Learn how to love? This is a new to me. I always thought that the foundation of success is persistence and a little bit of luck. So what have you learned?"

"I'm in the early stage. I have to learn a lot yet, and I'm learning. I think that not even the stars above me are higher than love. Love is more powerful than anything. I want to learn how to love, and anything what's below love, I don't need. Worries are below love.

Fears are below love. Grievance, hatred, jealousy, and judging are below love. All those negative aspects are nothing but a bottomless pit. From that pit it's impossible to see and feel the breadth and depth of life, and it's impossible to see the capabilities and full extent of it. Accepting all these negative aspects means sitting on the bottom of the pit and contemplating only a little piece of the sky above and, sometimes, the sun.

Imagine this, 'I want to go there – to the sky, but I cannot get out of the pit. What does it mean – I cannot?! I can fly! I have tremendous opportunities that I don't even know. I can fly, but what holds me back: the heavy load of pride, illusions, fears, insecurity, grievances, worries, jealousy, hatred, and negativity. That's what's holding me at the bottom and won't let me fly. Why do I need this load? I don't. I need to find the state of love.

No need to hurry to earn financial stability, social status, and so on. No need to strive hard for it. At the same time it's necessary to live fully every moment of life. A state of "here and now" has to be presented constantly. The state of love has to be here and now. Fill every moment of your life with love, and then you'll receive everything that you need! Everything will come when there are no hassles and an excessive desire to achieve, but there is peace and enjoyment of life as well. There has to be an action of course, but again, without fuss and hurry to live. There should be calm and purposeful movement.

I chose this way. When this state of love comes to me, it feels as if all the information of the universe is within me. Sometimes I can feel it periodically and sometimes it lasts longer. When the state of love comes to me, I don't need to think hard for the right answer. The answer comes immediately and it's always correct.

This is the state of knowledge and power. This is not a power over others, but a control over my own weaknesses. In this state there is no room for jealousy, envy, anger, frustration, and self-doubt. There is only love. At that point I don't need laws of society, morality, ethics, etiquette, or politeness.

This is a moment when *I am Love. I am generosity. I am compassion. I am tenderness and strength together. And if I am love, then all the doors are opened for me.* In the state of love there is no fear, no worries, and no illnesses.

The state of love creates such a space, such energy, and such atmosphere around a person that it begins to attract the good and repel the evil. So as a consequence dreams can come true.

I'm sure that *the key to harmony is to grow a state of love in your heart.* This state of love has to be carefully nurtured as a small sprout, so later it becomes mighty and strong.

When a person comes to this state, then a positive change occurs in the body. In this state, everything goes well and barriers disappear. Even if something unpleasant happens, it's easy to go through a hard time. When you reach this state, you become the owner of your life. Thus, it's useful to learn how not to be a slave of circumstances and how not to be a slave of negative emotions.

Often people ask, 'What is the meaning of life?' This question couldn't be answered for centuries by many seekers. I found the answer for myself. *The meaning of life is to learn how to love.* The sooner you learn it, the better.

Life is a school. *All the people and circumstances are teachers.* I chose to be a schoolboy in this school. Now, you have joined this school because you want to change your life for the better through the inner transformations. So, today you get lessons which in the beginning might look scary with their complexity, but tomorrow it would be a piece of cake to deal with them. If you studied diligently and were careful to follow the right principles, the day after they won't be noticeable to you. This is a great school and I've decided to accept it totally. I deny the position of the victim and I'm gradually becoming a master. Now it's your turn."

Slava's eyes were bright showing his great interest in my philosophy.

"It's awesome! I got it now! You're right. I was in the position of the victim. I choose to be a master. I'll reject anything that doesn't match my new position. Eventually, I'll become a master through developing a state of love within myself. Now, it can all come together. Now I know what to do and what it's for. You know, I'm in a great mood now, and even my liver doesn't hurt as much."

"There you go. This is the first result so don't stop and you will accomplish your goals. Think again about what we've discussed today. I'll get in touch with you in a couple of days, and I'll tell you what controlled stress is. I'll teach you a few techniques as well, so they will help you in your endeavors."

* * *

We shook hands, and I wished him luck. I remembered how one year ago I had a similar conversation thinking myself a sage. After I went through the effective procedures of removing the debris from my personality, I realized how little I had known before and how little yet I still know. I spoke to myself, "I will continue to grow past my imperfections and help others the best I can."

The next day Lenar delightedly shouted into the phone, "Alex, where is Slava? After meeting with you, he looks like another man. He is happy and smiling! Occasionally he leaves the office for some time to meditate. Did you give him a new brain? Alex, you're a magician! Mr. Voodoo, can I book an appointment with you?"

We both laughed heartily.

CHAPTER 11

I stayed in Moscow for three weeks and then I went home. During these three weeks, Lena and I had a pleasant time. I treated her with affection and care. Being young, tender, and beautiful, she respected and admired me. What more could a man dream of?

She was totally wonderful, but I knew it was not my way - to be with a young and sexy woman providing affection and care to her. I was drawn to the unknown places, new knowledge, and a spiritual search. At the same time, I had a strong feeling that I was not what she really needed. I clearly felt that this remarkable woman would soon meet her true prince. I did not want Lena to miss her happiness. Before I went home, I told her that I love her and she is free. Apparently, she was very upset and she did not want to hear about her true prince. She wanted to stay with me and to hear from me that I wanted to be with her forever. However, I could see further than my young goddess. I knew that she was still so inexperienced because she was attracted to the shiny candy wrapper she had first seen. She was not in love with me, but with the illusion she had created. I knew that she would have a beautiful future, but not with me. Because I loved Lena, I had to let her go.

We communicated through the Internet. She missed me, wanted to go see me, or asked me to come over to visit her. I felt great affection for her, but kept my distance because it was better to do so for her future. I did not see us together. I was just a transition in her life, and I had played my role.

Being in the state of love, I plunged into creativity and new projects. My senses did not deceived me, and after two and a half months Lena sent me a letter:

> *"Dear Alex, you cannot imagine how happy I am! You were right. I met a real prince. His name is Jack. He is a bit like you and thinks the same way. He has a lot of valuable knowledge and principles from his father. Although he isn't mature and wise like you're, but we can grow in maturity and wisdom together. Soon, we are going to get married. He is very kind and we love each other very much.*
>
> *I'm sorry that I was upset with you and my fate because you left me alone so suddenly when I had just started to realize what happiness was. Now I know what you meant. You don't want and cannot belong to anybody. You want to love unconditionally, and the only way to be with you is to accept your philosophy of life and accept you as you are.*
>
> *Now I understand how silly I was to feel upset with you. You appeared in my life as a mentor who taught me how to avoid mistakes and how to be happy. Because I've learned a lot from you, I know how to build a good relationship with my lover. I'm thankful to my fate and I'm grateful to you my darling. I know this letter won't hurt you and you will be happy for me.*
>
> *You are a model of a real man. I understand that you gave me everything you could. I understand that it's priceless. I respect you, and I love you, but now with a different love, as a sister.*
>
> *Yours truly, Lena."*

Everything was in order! My intuition hadn't failed me. I was happy for Lena. However, deep inside I was hurt in the weaker part of myself, but I quickly let it go and concentrated on my work. I thought, "I shouldn't just receive only, I have to give away too."

I really was genuinely happy for her. I was glad that I had done the right thing by giving Lena my love, thoughts on of life, perceptions of the world, and that I had let her go.

Then again I remembered the island and Ilistre. She gave me so much and left me without demanding anything. "You are a warrior of light" was Ilistre's goodbye. In that time I did not understand the meaning of those words.

In my understanding, a warrior was a fighter or a general. I thought in order to be a warrior I had to fight and destroy an enemy through the knowledge of tactics and strategies.

I used to be at war with my stupidity and shortcomings. I was simply firing the cannons at the sparrows not realizing that in a war with imperfections of the world wins fatigue, and I cannot be a winner because it's a fight with myself.

Later, I understood what Ilistre meant. To be a warrior of light meant spreading the good and dispelling the darkness with light. I had to spread the good abundantly. There was so much evil in the world and this process had not stopped yet. The creative human mind with scientific and technological progress had created so many means of destruction that if applied, would put an end to creativeness because there would be no human beings left.

The worldview of a sculptor is expressed best by a statue in the Vatican. The sculpture - a huge model of the Earth made of metal has been torn from within by another smaller ball. The sculpture is named "The Earth that has been destroyed by man." It is a symbol of how the evil, which continues to accumulate, can destroy our planet. So, it means in the presence of a narrow mind, it might lead to a self-destruction.

It's similar with cancer cells in the human body. With the intensive breeding they waste their energy on transportation. As a result, these flawed cells die together with the body which originally infected itself. However, a person who is infected with cancer is able to stop the dissemination of the disease in the body because an altered consciousness performs miracles in a human body. There are many examples of people being healed from the deadly diseases. I have my own experience when my body was healed after the expansion of my consciousness and clearing my mind of the "debris".

So, for me to be a Warrior of Light does not mean I had to fight with others as I did before. I had to consider all the mistakes that I had made, do good, and act competently and efficiently. It means to take part in the expansion of the consciousness of those who aspire to it. Of course we should start with the "renovation" of the inner world which I was able to accomplish during the year I met my mentors. It is proved that "the game is worth the candle." It happened that I was able to quickly understand, realize, and change many things in my life. However, not everybody who wants to change can and should do it the same way. Everyone has his or her own way. Nevertheless, I had the ability to tell others the basic techniques and principles of managing their lives.

When a person begins to "shine", the evil spirits become aggressive and they try to "put out" the light. Therefore, *who spreads the good and brings light must become strong and indestructible. The good must be strong!* What does it mean? What gives the strength? What takes the strength away? *Grievance, self-blame, self-dislike, pride, jealousy, and envy make a person weak.* These manifestations consume power and strength. *Kindness, compassion, generosity, the ability to control desires, ability to draw right conclusions and to follow them, gives strength.*

EPILOGUE

Well, I can take a breath and draw preliminary conclusions. My personal life is going well and the way I want. My business didn't fall apart during my absence, and survived. Moreover, now there are more opportunities for its expansion. The practice has shown that it can develop without my participation. Before my "vacation" I was able to establish it as a mechanism. Despite the fact that there were small issues, it resisted. My dream came true – now my business gives me freedom, but just a couple of years ago I was a slave to my own business.

Since I had more free time, I could do what my soul really desired. I followed Victor's example - I opened a private practice. "You can awaken the sleeping souls and you can light a fire in the hearts of others", Dianand had told me.

If I can, then I should do it. I cannot say that everybody has immediately run up to me to learn from my experience, conclusions, and observations, but still, some process has begun. This activity became like a springboard to my next step.

Later, I started renting a room where I began to teach. I'm glad that my knowledge is in demand and helps many people. I teach the different practices, meditations, and physical exercises that Ilistre had taught me as well as the philosophy of life that I had learned from other sages. The combination of philosophy, meditations, and physical exercises lead to a great effect in expanding the conciseness and strengthens health.

We are given everything by nature to be happy. By removing the obstacles, which interfere with the joy of life, we free our ability to become happy. We can share this knowledge with others, and by doing so, we can multiply the good in the world and we can make our beautiful planet purer.

When my work began to gain momentum, I established a connection with the people who wanted to leave me feedback. I began to receive a large number of different responses. Some people have said I was a genius which provoked the emergence of the celebrity disease and others said (through the third parties) I was mediocrity, contributing to the prevention of that disease.

I appreciated the opinions of both sides and I thought that they were useful in helping me control over my moods. I was grateful for the abundance of the contrasts because I became free from the influences of those people. I appreciate both sides because they help me to find my strength and virtue.

I recall with gratitude the guidance of the famous Russian poet Alexander Pushkin, "*... praise and slander accept with indifference and do not challenge a fool.*" I was thankful to the great poet for this guidance because it became a rule for me.

Furthermore, I noticed another interesting fact: the more I gained experience in a new field and increased my abilities, the more I realized how little I still knew and how little I could do, and there is a lot more to learn. Therefore, I don't miss the opportunity to obtain a new knowledge from other masters.

On the island I had learned that it is extremely important to control thoughts, words, and actions. The closer I get to the "normal mode" where thoughts are jumping uncontrollably, and when my body and desires are controlling me, the further away I am from the state of harmony and God. The further away I am from the divine and the closer I am to an animal. I was amazed that I comprehended the truth not among the civilized, technologically literate, and educated people, but among a tribe on an island.

In a modern society where one of the main virtues is to fill the mind with an encyclopedic knowledge and ignore the possibility of other knowledge and practices emphasizing different incorrect values in life. It is incredibly difficult to know the truth; it is difficult, but necessary!

Self-discovery opens a gate to the harmony. Harmony is happiness. Happiness is joy. Everyone wants to feel joy, so in order to fill life with joy there has to be an action – to leave the world alone with all its imperfections, stop blaming others, and step on the path of knowledge for the truth. It is a path of accumulation of wisdom and a path of disclosure of virtues from within. Now I know for sure, this is the only way. The power of love, compassion, and virtue create a wave that carries the travelers to their highest goals and to the implementation of dreams.

In conclusion I want to say ... no, I want to shout out to the entire universe, "**People! Love yourself and others! Love our planet! Do the good!**"

ABOUT THE TRANSLATOR

My name is **FEDOR MARKVARDT**. I was born in 1975 in Leningrad, Soviet Union. When I was a kid, I went to a daycare, pioneer camp, and school as many other Soviet children. At school I was a mischievous student receiving bad marks on all subjects, with the exception of physical education and handyman classes. My parents tried to make something useful of me, and I took guitar, piano, and balalaika classes, but to my parent's great disappointment I successfully forgot what I had learned. However, I reached a pretty good level in tennis.

During school years, I started smoking cigarettes, drinking alcohol, and doing drugs. Somehow, I finished the high school and went on to study in a university just to avoid compulsory military service that every guy over 18 had to do. I studied there for the whole three months before I was expelled for a fight with another student. Then, I decided to study to be a cook. I actually managed to finish the college, but I never had a chance to work in this field. The cooking skills proved very handy for me because I lived alone for two years after the rest of my family had emigrated to Canada.

From 1992 to 1995 four of my best friends died for different reasons. It was a very hard time for me. I thought I was literally losing my mind. Can you imagine, at the age of 21 I had many gray hairs on the back of my head. I was tremendously depressed.

In 1996 I got a Canadian visa, so I emigrated to Canada and lived with my parents and sister in Edmonton. I did a number of odd jobs: delivered newspapers and pizza, drove a truck delivering heavy doors and windows for new homes, and worked in a construction company. One day I seriously hurt my shoulder lifting a 400 pounds linoleum piece. I took some days off and then realized that I need a different job, so I quit.

In 1997 my life changed dramatically - I broke my neck and became a quadriplegic. It was a huge shock for me, my family, and friends. Doctors did not want to operate on my neck for the first three days after the injury because they thought I was not going to make it. But I survived. I guess God saved me and gave me another opportunity to understand something and do something useful.

First few years I basically spent in bed because I was weak physically and spiritually. I was afraid to go out, talk to people, and had no interest in doing anything. I was even thinking of committing suicide. Because my mother was cheering me up, I overcame those kinds of thoughts.

In 1999 my mother died, so it was an even more difficult time for everybody in my family. I was recovering from her passing and the shock of being paralyzed very slowly, but with a great support from my sister, my Dad, and my friends, I managed to find some confidence. I also received great support from the Canadian Paraplegic Association and especially from Guy Coulombe. Later, I was persuaded by Margaret Conquest, who also worked in the CPA office, to go back to school and a few months later I decided to take a computer course in a college. Later, I took several other courses including English and Spanish languages.

Then, I took seven brilliant courses for health and spirit improvement and became a different person. I got rid of my fears, grievances, complaints, and finally I started to enjoy my life. While I attended the trainings of Alexander Markitanov, I gained the knowledge of ancient sages. I have been collecting this invaluable information from different sources for many years. I noticed that my knowledge aroused interest in other people and therefore I decided to conduct webinars online and do trainings on skype. I worked out a careful plan on how to help people cultivate their latent capabilities.

In 2014 my father died. Because of what I learned at Alexander Markitanov trainings about how to face a loss, I recovered pretty quickly, but still it was a tremendous shock for me. I couldn't work on my website and particularly on finishing and publishing the book "Stairway to a dream". Without a doubt, the death of the closest person is the hardest thing to deal with, but if a person knows the 'literate view' of how to face a loss, as described in the book, it is much easier to overcome the tough time.

I've been working hard to improve myself for quite a long time and now I can say, "I'm proud of myself!" Looking back, I see what a tremendous journey I have made. But I don't look back for too long. Every day I work with my weaknesses and improve the technique to expand the consciousness. I communicate with my mentors often and I continue to develop new techniques. I never lose hope to walk again and to have my own family.

Every day life throws at us different lessons to be learned, but not everyone knows how to study in the school of life. I aspire to help other people learn more about these lessons and to acquire the skills that could improve their lives. Because of stereotypes, preconceptions, erroneous believes and fears that people might have, they stay "blind" and it interferes with the wise perception of the world. As individuals develop toxic ideas, they do harmful things and quite often pay a high price for them. However, many people are too afraid or too weak to change. What can I say? It's their choice and they construct their life for themselves. By the end of the life they might regret what they didn't do rather than what they did.

If it so happens that you are often in a bad mood, having problems at school, work, or home and you want to change your life for the better – I'd love to help you! I cannot solve your problems for you, but I will work with you to help you become confident, strong, and successful, so you could enjoy your life 100%! Most importantly, if you really want to accomplish something, you must make an effort yourself.

I was able to quit bad habits, come out of depression, and to start enjoying the life fully. I sincerely wish to see you healthy, strong, and happy!

Visit my website: http://www.becomejoyful.com/

CPSIA information can be obtained
at www.ICGtesting.com
Printed in the USA
LVOW12s0510310316
481500LV00003BA/24/P

9 781460 284537